SAS® Viya®
The R Perspective

Yue Qi
Kevin D. Smith
Xiangxiang Meng

The correct bibliographic citation for this manual is as follows: Qi, Yue, Kevin D. Smith, and Xiangxiang Meng. 2018. *SAS® Viya®: The R Perspective*. Cary, NC: SAS Institute Inc.

SAS® Viya®: The R Perspective

Contents

About This Book

What Does This Book Cover?

This book is an introduction to using the R client on the SAS Viya platform. SAS Viya is a high-performance, fault-tolerant analytics architecture that can be deployed on both public and private cloud infrastructures. Although SAS Viya can be used by various SAS applications, it also enables you to access analytic methods from SAS, R, Python, Lua, and Java, as well as through a REST interface using HTTP or HTTPS. Of course, in this book we focus on the perspective of SAS Viya from R.

SAS Viya consists of multiple components. The central piece of this ecosystem is SAS Cloud Analytic Services (CAS). CAS is the cloud-based server that all clients communicate with to run analytical methods. The R client is used to drive the CAS component directly using objects and constructs that are familiar to R programmers.

We assume that you have some knowledge about R before you approach the topics in this book. We do not assume any knowledge of CAS itself. However, you must have a CAS server that is set up and is running in order to execute the examples in this book.

The chapters in the first part of the book cover topics from the installation of R to the basics of connecting, loading data, and getting simple analyses from CAS. Depending on your familiarity with R, after reading the "Ten-Minute Guide to Using CAS from R," you might feel comfortable enough to jump to the chapters later in the book that are dedicated to statistical methods. However, the chapters in the middle of the book cover more detailed information about working with CAS, such as constructing action calls to CAS and processing the results, error handling, managing your data in CAS, and using object interfaces to CAS actions and CAS data tables. Finally, the last chapter about advanced topics covers features and workflows that you might want to take advantage of when you are more experienced with the R client.

This book covers topics that are useful to complete beginners, as well as to experienced CAS users. Its examples extend from creating connections to CAS to simple statistics and machine learning. The book is also useful as a desktop reference.

Is This Book for You?

If you are using the SAS Viya platform in your work and you want to access analytics from SAS Cloud Analytic Services (CAS) using R, then this book is a great starting point. You'll learn about general CAS workflows, as well as the R client that is used to communicate with CAS.

What Are the Prerequisites for This Book?

Some R experience is definitely helpful while reading this book. If you do not know R, there is a multitude of resources on the internet for learning R. The later chapters in the book cover data analysis and modeling topics. Although the examples provide step-by-step code walk-throughs, some training about these topics beforehand is helpful.

What Should You Know about the Examples?

This book includes tutorials for you to follow to gain hands-on experience with SAS.

Software Used to Develop the Book's Content

This book was written using Version 1.3.0 of the SAS Scripting Wrapper for Analytics Transfer (SWAT) package for R. SAS Viya 3.3 was used. Various R resources and packages were used as well. SWAT works with many versions of these packages. The URLs of SWAT and other resources are shown as follows:

SAS Viya
www.sas.com/en_us/software/viya.html

SAS Scripting Wrapper for Analytics Transfer (SWAT) – R client to CAS
github.com/sassoftware/R-swat **(GitHub repository)**

R
https://www.r-project.org/

RStudio – an integrated development environment (IDE) for R
https://www.rstudio.com/

Example Code and Data

You can access the example code and data for this book by going to the author page at
https://support.sas.com/authors or on GitHub at: https://github.com/sassoftware/sas-viya-the-R-perspective.

We Want to Hear from You

SAS Press books are written *by* SAS Users *for* SAS Users. We welcome your participation in their development and your feedback on SAS Press books that you are using. Please visit sas.com/books to do the following:

- Sign up to review a book
- Recommend a topic
- Request information on how to become a SAS Press author
- Provide feedback on a book

Do you have questions about a SAS Press book that you are reading? Contact the authors through saspress@sas.com or https://support.sas.com/author_feedback.

SAS has many resources to help you find answers and expand your knowledge. If you need additional help, see our list of resources: sas.com/books.

About These Authors

 Yue Qi, PhD, is a staff scientist at SAS. He works on automated and adaptive machine learning pipelines, deep learning models on unstructured data, interactive data visualization, and open-source language integration. He has extensive experience in applying these technologies to develop analytics products, build successful models on big data for customers, and help customers solve their most challenging business problems, especially in the finance industry.

 Kevin D. Smith has been a software developer at SAS since 1997. He began his career in the development of PROC TEMPLATE and other underlying ODS technologies, including authoring two books on the subjects. He is now heavily involved in client-side work on the SAS® Viya® platform. This includes development of the R, Python, and Lua SWAT packages, as well as higher-level packages built on top of the foundation created by SWAT.

 Xiangxiang Meng, PhD, is a Senior Product Manager at SAS. The current focus of his work is on SAS® Visual Statistics, deep learning, reinforcement learning, the Python interface to SAS® Viya®, and other new product initiatives. Previously, Xiangxiang worked on SAS® LASR™ Analytic Server, SAS® In-Memory Statistics for Hadoop, SAS Recommendation Systems, and SAS® Enterprise Miner™. His research interests include deep learning and reinforcement learning, automated and cognitive pipelines for business intelligence and machine learning, and parallelization of machine learning algorithms on distributed data. Xiangxiang received his PhD and MS from the University of Cincinnati.

Learn more about these authors by visiting their author pages, where you can download free book excerpts, access example code and data, read the latest reviews, get updates, and more:
http://support.sas.com/Qi
http://support.sas.com/Smith
http://support.sas.com/Meng

Chapter 1: Installing R, SAS SWAT, and CAS

Introduction

There are four primary pieces of software that must be installed in order to use SAS Cloud Analytic Services (CAS) from R:

- 64-bit version of R 3.1.0 or later
- the SAS SWAT R package
- the dplyr, http, and jsonlite packages. These packages have additional dependencies that are automatically installed from CRAN when you run the `install.packages()` function.
- the SAS CAS server

We cover the recommended ways to install each piece of software in this chapter.

Installing R

The R packages that are used to connect to SAS Viya have a minimum requirement of R 3.1.0. If you are not familiar with R or if you don't have a version preference, we recommend that you use the most recent release of R. You can download R at https://cran.r-project.org/.

After you have installed R, the next step is to install the SAS SWAT package.

Installing SAS SWAT

The SAS SWAT package is the R package created by SAS that is used to connect to SAS Viya. SWAT stands for SAS Scripting Wrapper for Analytics Transfer. It includes two interfaces to SAS Viya: 1) a natively compiled client for binary communication, and 2) a pure R REST client for HTTP-based connections. Support for the different protocols varies based on the platform that is used. So, you'll have to check the downloads on the GitHub project to find out what is available for your platform.

To install SWAT, use the standard R installation function `install.packages()`. The SWAT installers are located at GitHub in the `r-swat` project of the `sassoftware` account. The available releases are listed at the following link:

https://github.com/sassoftware/r-swat/releases

After downloading the package, you can install SWAT using a command similar to the following:

```
R CMD INSTALL R-swat-X.X.X-platform.tar.gz
```

where *X.X.X* is the version number and *platform* is the platform that you are installing on.

You can also install the SWAT package from the URL directly using the following code in R:

```
# Make sure prerequisites are installed

> install.packages('httr')

> install.packages('jsonlite')

> install.packages('dplyr')

> install.packages('https://github.com/sassoftware/R-
swat/releases/download/vX.X.X/R-swat-X.X.X-platform.tar.gz',repos=NULL,
type='file')
```

For example, you can use the following R code to install SWAT version 1.3.0 on your Linux 64 machine:

```
> install.packages('https://github.com/sassoftware/R-
swat/releases/download/1.3.0/R-swat-1.3.0-linux64.tar.gz.tgz', repos=NULL,
type='file')
```

If you are on a platform where only the REST interface is available, you can use the REST installer for that platform. For example, you can use the following R code to install version 1.3.0 on a OS X machine:

```
> install.packages('https://github.com/sassoftware/R-
swat/releases/download/1.3.0/R-swat-1.3.0-osx-REST-only.tar.gz', repos=NULL,
type='file')
```

If your platform isn't in the list of available packages, you can install using the source code URL on the releases page instead, but you are restricted to using the REST interface over HTTP or HTTPS.

```
 > install.packages('https://github.com/sassoftware/R-
swat/archive/vX.X.X.tar.gz', repos=NULL, type='file')
```

After SWAT is installed, you should be able to run the following command in R to load the SWAT package:

```
> library('swat')
```

You can submit the preceding code in plain RGui or RStudio. You can also use the popular Jupyter notebook with the R kernel installed, which was previously known as the IPython notebook. Jupyter is most commonly used within a web browser. It can be launched with the jupyter notebook command at the command line.

In this book, we primarily show plain text output using RStudio. However, all of the code from this book is also available in the form of Jupyter notebooks here:

https://github.com/sassoftware/sas-viya-the-R-perspective

Now that we have installed R and SWAT, the last thing we need is a CAS server.

Installing CAS

The installation of SAS Cloud Analytic Services (CAS) is beyond the scope of this book. Installation on your own server requires a CAS software license and system administrator privileges. Contact your system administrator about installing, configuring, and running CAS.

Making Your First Connection

With all of the pieces in place, let's make a test connection just to verify that everything is working. From R, you should be able to run the following commands:

```
> library('swat')
> conn <- CAS('server-name.mycompany.com', port = port-number,
              username = 'userid', password = 'password',
              protocol = 'http')
> cas.builtins.serverStatus(conn)
> cas.terminate(conn)
```

Where

- *server-name.mycompany.com* is the name or IP address of your CAS server,

- *port-number* is the port number that CAS is listening to,

- *userid* is your CAS user ID,

- *password* is your CAS password.

The cas.builtins.serverStatus function returns information about the CAS grid that you are connected to, and the cas.terminate function closes the connection. If the commands run successfully, then you are ready to move on. If not, you'll have to do some troubleshooting before you continue.

Conclusion

At this point, you should have R and the SWAT package installed, and you should have a running CAS server. In the next chapter, we'll give a summary of what it's like to use CAS from R. Then, we'll dig into the chapters that go into the details of each aspect of SWAT.

Chapter 2: The Ten-Minute Guide to Using CAS from R

If you are already familiar with R, have a running CAS server, and just can't wait to get started, we've written this chapter just for you. This chapter is a very quick summary of what you can do with CAS from R. We don't provide a lot of explanation of the examples; that comes in the later chapters. This chapter is here for those who want to dive in and work through the details in the rest of the book as needed.

Loading SWAT and Getting Connected

The only thing that you need to know about the CAS server in order to get connected is the host name, the port number, your user name, and your password. The last two items might even be optional if you are using an Authinfo file, which is explained in detail in Chapter 3. The SWAT package contains the CAS class that is used to talk to the server. The arguments to the CAS class are host name, port, user name, and password, in that order.[1] Note that you can use the REST interface by specifying the HTTP port that is specified by the CAS server. The CAS class can auto detect the port type for the standard CAS port and HTTP. However, if you use HTTPS, you must specify protocol='https' as a keyword argument when you start a CAS connection. You can also specify 'cas' or 'http' to explicitly override auto detection.

```
> library('swat')
```

```
SWAT 0.1.3
```

```
> conn <- CAS('server-name.mycompany.com', 8777, 'username', 'password')
```

```
Connecting to CAS and generating CAS action functions for loaded action sets...
To generate the functions with signatures (for tab completion), add
'genActSyntax=TRUE' to your connection parms.
```

When you connect to CAS, it creates a session on the server. By default, all resources (CAS actions, data tables, options, and so on) are available only to that session. Some resources can be promoted to a global scope, which we discuss later in the book.

To see what CAS actions are available, use the cas.builtins.help method on the CAS connection object, which calls the help action in builtins action set on the CAS server.

```
> out <- cas.builtins.help(conn)
```

```
NOTE: Available Action Sets and Actions:
NOTE:    accessControl
NOTE:       assumeRole - Assumes a role
NOTE:       dropRole - Relinquishes a role
NOTE:       showRolesIn - Shows the currently active role
NOTE:       showRolesAllowed - Shows the roles that a user is a member
                    of
NOTE:       isInRole - Shows whether a role is assumed
NOTE:       isAuthorized - Shows whether access is authorized
NOTE:       isAuthorizedActions - Shows whether access is authorized to
                    actions
NOTE:       isAuthorizedTables - Shows whether access is authorized to
                    tables
NOTE:       isAuthorizedColumns - Shows whether access is authorized to
                    columns
NOTE:       listAllPrincipals - Lists all principals that have explicit
                    access controls
NOTE:       whatIsEffective - Lists effective access and explanations
                    (Origins)
...
NOTE:       partition - Partitions a table
NOTE:       shuffle - Randomly shuffles a table
NOTE:       recordCount - Shows the number of rows in a Cloud Analytic
                    Services table
NOTE:       loadDataSource - Loads one or more data source interfaces
NOTE:       update - Updates rows in a table
```

The return values from all actions are in the form of the R list class. To see a list of names of all of the list members, use the names() function just as you would with any R list. In this case, the object names correspond to the names of the CAS action sets.

```
> names(out)
 [1] "accessControl"  "builtins"      "configuration"
 [4] "dataPreprocess" "dataStep"      "percentile"
 [7] "search"         "session"       "sessionProp"
[10] "simple"         "table"
```

Printing the contents of the return value shows all of the top-level list members as sections. The builtins.help action returns the information about each action set in a table. These tables are stored in the output as casDataFrames.

```
> out
$accessControl
```

	Name	Description
1	assumeRole	Assumes a role
2	dropRole	Relinquishes a role
3	showRolesIn	Shows the currently active role
4	showRolesAllowed	Shows the roles that a user is a member of
5	isInRole	Shows whether a role is assumed

	Name	Description
6	isAuthorized	Shows whether access is authorized
7	isAuthorizedActions	Shows whether access is authorized to actions
8	isAuthorizedTables	Shows whether access is authorized to tables
9	isAuthorizedColumns	Shows whether access is authorized to columns
10	listAllPrincipals	Lists all principals that have explicit access controls
…	…	…
20	partition	Partitions a table
21	shuffle	Randomly shuffles a table
22	recordCount	Shows the number of rows in a Cloud Analytic Services table
23	loadDataSource	Loads one or more data source interfaces
24	update	Updates rows in a table

Since the output is based on R's list object, you can access each list member individually as well.

```
> out$builtins
```

	Name	Description
1	addNode	Adds a machine to the server
2	removeNode	Remove one or more machines from the server
3	help	Shows the parameters for an action or lists all available actions
4	listNodes	Shows the host names used by the server
5	loadActionSet	Loads an action set for use in this session
6	installActionSet	Loads an action set in new sessions automatically
7	log	Shows and modifies logging levels
8	queryActionSet	Shows whether an action set is loaded
9	queryName	Checks whether a name is an action or action set name
10	reflect	Shows detailed parameter information for an action or all actions in an action set
11	serverStatus	Shows the status of the server
12	about	Shows the status of the server
13	shutdown	Shuts down the server
14	userInfo	Shows the user information for your connection
15	actionSetInfo	Shows the build information from loaded action sets
16	history	Shows the actions that were run in this session
17	casCommon	Provides parameters that are common to many actions
18	ping	Sends a single request to the server to confirm that the connection is working
19	echo	Prints the supplied parameters to the client log
20	modifyQueue	Modifies the action response queue settings
21	getLicenseInfo	Shows the license information for a SAS product
22	refreshLicense	Refresh SAS license information from a file
23	httpAddress	Shows the HTTP address for the server monitor

Running CAS Actions

Just like the builtins.help action, all of the actions are available as R functions. You need to specify the fully qualified name of the action, which includes both the action set name and the action name. For example, the userInfo action is contained in the builtins action set. To call it, you have to use the full name cas.builtins.userinfo. Note that both the action set name and the action name are always written in camelCase.

For example, the userInfo action is called as follows.

```
> cas.builtins.userInfo(conn)
```

```
$userInfo
$userInfo$anonymous
[1] FALSE

$userInfo$groups
$userInfo$groups[[1]]
[1] "users"


$userInfo$hostAccount
[1] TRUE

$userInfo$providedName
[1] "username"

$userInfo$providerName
[1] "Active Directory"

$userInfo$uniqueId
[1] "username"

$userInfo$userId
[1] "username"
```

The result this time is still a list object, and the contents of that object is another list (userInfo) that contains information about your user account. Although all actions return a list object, there are no strict rules about what member names and values are in that object. The returned values are determined by the action and they vary depending on the type of information returned. Analytic actions typically return one or more casDataFrames.

Loading Data

The easiest way to load data into a CAS server is by using the as.casTable() function. This function uploads the data from an R data.frame to a CAS table. We use the classic Iris data set in the following data-loading example.

```
> iris_ct <- as.casTable(conn,iris)
> attributes(iris_ct)
```

```
$conn
CAS(hostname=server-name.mycompany.com, port=8777, username=username,
session=60c6e0fc-d690-ea48-9dbc-9692e7205455, protocol=http)

$tname
[1] "iris"

$caslib
[1] ""

$where
[1] ""

$orderby
[1] ""

$groupby
[1] ""

$gbmode
[1] ""

$computedOnDemand
[1] FALSE

$computedVars
[1] ""

$computedVarsProgram
[1] ""

$names
[1] "Sepal.Length" "Sepal.Width"  "Petal.Length" "Petal.Width"  "Species"

$class
[1] "CASTable"
attr(,"package")
[1] "swat"
```

The output from the as.casTable() function is a CASTable object. The CASTable object contains the connection information, name of the created table, the caslib that the table was created in, and other information. The CASTable objects also support many of the operations that are defined by R data.frame so that you can operate on them as if they were local data.[2]

You can use actions such as tableInfo and columnInfo in the table action set to access general information about the table itself and its columns.

```
# Call the tableInfo action on the CASTable object.
> cas.table.tableInfo(conn)
```

```
  Name Rows Columns Encoding CreateTimeFormatted   ModTimeFormatted
1 IRIS  150       5    utf-8  17Apr2017:02:17:40 17Apr2017:02:17:40
  JavaCharSet CreateTime    ModTime Global Repeated View SourceName
1        UTF8 1808014660 1808014660      0        0    0
  SourceCaslib Compressed   Creator Modifier
1                       0  username
```

```
# Call the columninfo action on the CASTable.
> cas.table.columnInfo(iris_ct)
$ColumnInfo
        Column ID    Type RawLength FormattedLength NFL NFD
1 Sepal.Length  1  double         8              12   0   0
2  Sepal.Width  2  double         8              12   0   0
3 Petal.Length  3  double         8              12   0   0
4  Petal.Width  4  double         8              12   0   0
5      Species  5 varchar        10              10   0   0
```

Now that we have some data, let's run some more interesting CAS actions on it.

Executing Actions on CAS Tables

The simple action set that comes with CAS contains some basic analytic actions. Let's run the summary action from the simple action set on our CAS table.

```
> summ <- cas.simple.summary(iris_ct)
> summ
```

```
$Summary
        Column Min Max   N NMiss     Mean     Sum       Std
1 Sepal.Length 4.3 7.9 150     0 5.843333 876.5 0.8280661
2  Sepal.Width 2.0 4.4 150     0 3.057333 458.6 0.4358663
3 Petal.Length 1.0 6.9 150     0 3.758000 563.7 1.7652982
4  Petal.Width 0.1 2.5 150     0 1.199333 179.9 0.7622377
       StdErr       Var     USS       CSS       CV   TValue
1 0.06761132 0.6856935 5223.85 102.16833 14.17113 86.42537
2 0.03558833 0.1899794 1430.40  28.30693 14.25642 85.90830
3 0.14413600 3.1162779 2582.71 464.32540 46.97441 26.07260
4 0.06223645 0.5810063  302.33  86.56993 63.55511 19.27060
         ProbT
1 3.331256e-129
2 8.004458e-129
3  2.166017e-57
4  2.659021e-42
```

The summary action displays summary statistics in a form that is familiar to SAS users. If you want them in a form that is similar to what R users are used to, you can use the summary() method (just like on R data.frame objects).

```
> summary(iris_ct)
```

```
  Sepal.Length    Sepal.Width     Petal.Length    Petal.Width
 Min.   :4.300   Min.   :2.000   Min.   :1.000   Min.   :0.100
 1st Qu.:5.100   1st Qu.:2.800   1st Qu.:1.600   1st Qu.:0.300
 Median :5.800   Median :3.000   Median :4.350   Median :1.300
 Mean   :5.843   Mean   :3.057   Mean   :3.758   Mean   :1.199
 3rd Qu.:6.400   3rd Qu.:3.300   3rd Qu.:5.100   3rd Qu.:1.800
 Max.   :7.900   Max.   :4.400   Max.   :6.900   Max.   :2.500
       Species
 setosa    :50
 versicolor:50
 virginica :50
```

Note that when you call the summary() function on a CASTable object, it calls various CAS actions in the background to do the calculations. This includes the cas.table.columnInfo, cas.simple.summary, cas.percentile.percentile, and cas.fedsql.execDirect actions. The output of those actions is combined into a data.frame in the same form that the real R summary() function returns. This enables you to use CASTable objects and R data.frame objects interchangeably in your workflow to work on the result tables from CAS.

Data Visualization

Since the tables that come back from the CAS server are subclasses of an R data.frame, you can do anything to them that works on a data.frame. You can plot the results of your actions using the plot function or use them as input to more advanced packages, such as ggplot2, which are covered in more detail in a later section.

The following example uses the plot method to download the data set and plot it using the default options.[3]

```
> plot(iris_ct$Sepal.Length, iris_ct$Sepal.Width)
```

The output that is created by the plot function is shown in Figure 2.1.

Figure 2.1: Scatter Plot of Sepal.Width versus Sepal.Length

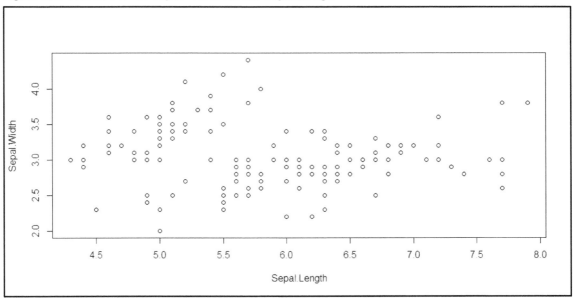

Closing the Connection

As with any network or file resource in R, you should close your CAS connections when you are finished. They time out and disappear eventually if left open, but it's always a good idea to clean them up explicitly.

```
> cas.terminate(conn)
```

Conclusion

Hopefully, this ten-minute guide was enough to give you an idea of the basic workflow and capabilities of the R CAS client. In the following chapters, we dig deeper into the details of the R CAS client and how to blend the power of SAS analytics with the tools that are available in the R environment.

---

[1] Later in the book, we show you how to store your password so that you do not need to specify it in your programs.

[2] However, until you explicitly fetch the data or call a function that returns data from the table (such as head or tail), all operations are simply combined on the client side (essentially creating a client-side view) until they are needed for the call to the CAS server for data.

[3] To prevent downloading very large data sets to the client, a maximum of only 10,000 rows can be randomly sampled and downloaded when the data set has more than 10,000 rows.

Chapter 3: The Fundamentals of Using R with CAS

The SAS SWAT package includes a programming interface to CAS, as well as utilities to handle results, format data values, and upload data to CAS. We have already covered the installation of SWAT in an earlier chapter, so let's jump right into connecting to CAS.

There is a lot of detailed information about parameter structures, error handling, and authentication in this chapter. If you feel like you are getting bogged down, you can always skim over this chapter and come back to it later when you need more formal information about programming using the CAS interface.

Connecting to CAS

In order to connect to a CAS host, you need some form of authentication. There are various authentication mechanisms that you can use with CAS. The different forms of authentication are beyond the scope of this book, so we use user name and password authentication in all of our examples. This form of authentication assumes that you have a login account on the CAS server that you are connecting to. The disadvantage of using a user name and password is that you typically include your password in the source code. However, Authinfo is a solution to this problem, so we'll show you how to store authentication information using Authinfo as well.

Let's make a connection to CAS using an explicit user name and a password. For this example, we use an R shell.

The first thing that you need to do after starting R is to import the SWAT package. This package contains a function called CAS that is the primary interface to your CAS server. It requires at least two arguments: the CAS host name or IP address, and the port number that CAS is running on[1]. Since we use user name and password authentication, we must specify them as the next two arguments. If there are no connection errors, you should now have an open CAS session that is referred to by the conn variable.

```
> library('swat')

> conn <- CAS('server-name.mycompany.com', 8777,
                    'username', 'password', protocol = 'http')

> conn
```

```
CAS(hostname=server-name.mycompany.com, port=8777, username=username,
session=ffee6422-96b9-484f-a868-03505b32098, protocol=http)
```

We use the http protocol in this example because it is available on all operating systems. As you can see in the output above, the string representation of the CAS object is displayed. You see that it echoes the host name, the port, the user name, and several fields that were not specified. The session field is created once the session is created. The session value contains a unique ID that can be used to make other connections to that same session.

We mentioned using Authinfo rather than specifying your user name and password explicitly in your programs. The Authinfo specification is based on an older file format called Netrc. Netrc was used by FTP programs to store user names and passwords so that you did not have to enter authentication information manually. Authinfo works the same way, but it adds a few extensions.

The basic format of an Authinfo file follows: (The format occupies two lines to enhance readability.)

```
host server-name.mycompany.com port 8777
     user username password password
```

Where *server-name.mycompany.com* is the host name of your CAS server (an IP address can also be used), *8777* is the port number of the CAS server, *username* is your user ID on that machine, and *password* is your password on that machine. If you don't specify a port number, the same user name and password are used on any port on that machine. Each CAS host requires a separate host definition line. In addition, the host name must match exactly what is specified in the CAS constructor. There is no DNS name expansion if you use a shortened name, such as *server-name*.

By default, the Authinfo file is accessed from your home directory under the name .authinfo (on Windows, the Authinfo file has to be named _authinfo.). It also must have permissions that are set up so that only the owner can read it. This is done using the following command on Linux:

```
chmod 0600 ~/.authinfo
```

On Windows, the file permissions should be set so that the file is not readable by the Everyone group. Once that file is in place and has the correct permissions, you should be able to make a connection to CAS without specifying your user name and password explicitly.

```
> library('swat')

> conn <- CAS('server-name.mycompany.com', 8777, protocol = 'http')

> conn
```

```
CAS(hostname=server-name.mycompany.com, port=8777,
username=username, session=ffee6422-96b9-484f-a868-03505b320987,
protocol=http)
```

After connecting to CAS, we can continue to a more interesting topic: running CAS actions.

Running CAS Actions

In the previous section, we made a connection to CAS, but didn't explicitly perform any actions. However, after the connection was made, many actions were performed to obtain information about the server and which resources are available to the CAS installation. One of the things queried for is information about the currently loaded *action sets*. An action set is a logical grouping of actions that perform related functions. Actions can do various things, such as return information about the server setup, load data, and perform advanced analytics. To see what action sets and actions are already loaded, you can call the builtins.help action on the CAS object that we previously created.

```
> out <- cas.builtins.help(conn)
```

```
NOTE: Available Action Sets and Actions:
NOTE:    accessControl
NOTE:       assumeRole - Assumes a role
NOTE:       dropRole - Relinquishes a role
NOTE:       showRolesIn - Shows the currently active role
NOTE:       showRolesAllowed - Shows the roles that a user is
                          a member of
NOTE:       isInRole - Shows whether a role is assumed
NOTE:       isAuthorized - Shows whether access is authorized
NOTE:       isAuthorizedActions - Shows whether access is
                 authorized to actions
NOTE:       isAuthorizedTables - Shows whether access is authorized
                                 to tables
NOTE:       isAuthorizedColumns - Shows whether access is authorized
                                  to columns
NOTE:       listAllPrincipals - Lists all principals that have
                                explicit access controls
NOTE:       whatIsEffective - Lists effective access and
                              explanations (Origins)
NOTE:       listAcsData - Lists access controls for caslibs, tables,
                          and columns
NOTE:       listAcsActionSet - Lists access controls for an action
                               or action set
NOTE:       repAllAcsCaslib - Replaces all access controls for
                              a caslib
NOTE:       repAllAcsTable - Replaces all access controls for a table
NOTE:       repAllAcsColumn - Replaces all access controls for
                              a column
NOTE:       repAllAcsActionSet - Replaces all access controls for
                                 an action set
NOTE:       repAllAcsAction - Replaces all access controls for
                              an action
NOTE:       updSomeAcsCaslib - Adds, deletes, and modifies some
                               access controls for a caslib
NOTE:       updSomeAcsTable - Adds, deletes, and modifies some
                              access controls for a table
NOTE:       updSomeAcsColumn - Adds, deletes, and modifies some
                               access controls for a column
NOTE:       updSomeAcsActionSet - Adds, deletes, and modifies some
                                  access controls for an action set
NOTE:       updSomeAcsAction - Adds, deletes, and modifies some
                               access controls for an action
NOTE:       remAllAcsData - Removes all access controls for a
                            caslib, table, or column
```

This prints out a listing of all of the loaded action sets and the actions within them. It also returns a list of casDataFrame structures that contain the action set information in tabular form. The results of CAS actions are discussed later in this chapter.

The builtins.help action takes arguments that specify which action sets and actions you want information about. To display help for an action set, use the actionset keyword parameter. The following code displays the help content for the builtins action set.

```
> out <- cas.builtins.help(conn, actionset = 'builtins')
```

```
NOTE: Information for action set 'builtins':
NOTE:     builtins
NOTE:        addNode - Adds a machine to the server
NOTE:        removeNode - Remove one or more machines from the server
NOTE:        help - Shows the parameters for an action or lists all
                   available actions
NOTE:        listNodes - Shows the host names used by the server
NOTE:        loadActionSet - Loads an action set for use in this
                   session
NOTE:        installActionSet - Loads an action set in new sessions
                   automatically
NOTE:        log - Shows and modifies logging levels
NOTE:        queryActionSet - Shows whether an action set is loaded
NOTE:        queryName - Checks whether a name is an action or
                   action set name
NOTE:        reflect - Shows detailed parameter information for an
                   action or all actions in an action set
NOTE:        serverStatus - Shows the status of the server
NOTE:        about - Shows the status of the server
NOTE:        shutdown - Shuts down the server
NOTE:        userInfo - Shows the user information for your connection
NOTE:        actionSetInfo - Shows the build information from loaded
                   action sets
NOTE:        history - Shows the actions that were run in this session
NOTE:        casCommon - Provides parameters that are common to many
                   actions
NOTE:        ping - Sends a single request to the server to confirm
                   that the connection is working
NOTE:        echo - Prints the supplied parameters to the client log
NOTE:        modifyQueue - Modifies the action response queue settings
NOTE:        getLicenseInfo - Shows the license information for a
                   SAS product
NOTE:        refreshLicense - Refresh SAS license information from
                   a file
NOTE:        httpAddress - Shows the HTTP address for the server
                   monitor
```

Notice that "help" is one of the actions in the builtins action set. To display the help for an action, use the action keyword argument. You can display the help for the "help" action as follows:

```
> out <- cas.builtins.help(conn, action = 'help')
```

```
NOTE: Information for action 'builtins.help':
NOTE: The following parameters are accepted.
     Default values are shown.
NOTE:     string action=NULL,
```

```
NOTE:      specifies the name of the action for which you want help.
           The name can be in the form 'actionSetName.actionName' or
           just 'actionName'.
NOTE:   string actionSet=NULL,
NOTE:      specifies the name of the action set for which you
           want help. This parameter is ignored if the action
           parameter is specified.
NOTE:   boolean verbose=true
NOTE:      when set to True, provides more detail for each parameter.
```

Looking at the printed notes, you can see that the builtins.help action takes the parameters actionset, action, and verbose. We have previously seen the actionset and action parameters. The verbose parameter is enabled, which means that you will get a full description of all of the parameters of the action. You can suppress the parameter descriptions by specifying verbose=FALSE, as follows:

```
> out <- cas.builtins.help(conn, action = 'help', verbose = FALSE)
```

```
NOTE: Information for action 'builtins.help':
NOTE: The following parameters are accepted.
      Default values are shown.
NOTE:   string action=NULL,
NOTE:   string actionSet=NULL,
NOTE:   boolean verbose=true
```

In addition to the help system that is provided by CAS, the SWAT package also enables you to access the information about some functions in SWAT using mechanisms that are supplied by R. R supplies the help function to display information about R objects. This same help function can be used to display information about some functions in SWAT. Note that it does not work for CAS actions and action sets. We have been using the builtins.help action on our CAS object. Let's see what the R help function displays.

```
> help(as.casTable)
```

An R documentation page shows up that displays the information on the as.casTable function. This function uploads an R data.frame to CAS and returns a CASTable object. A complete list of functions, from which you can use the R help function to get the help page, can be found in the API document on https://developer.sas.com/guides/r.html.

You can also use a question mark followed by the function name.

```
> ?as.casTable
```

Due to the fact that R packages require the documentation to be created when the package is compiled, it is not possible to generate action documentation at run time in order to take advantage of the '?' operator in R for displaying help. Therefore, you need to use either the builtins.help action or the CAS actions and action sets in the SAS Help Center. You can also simply do an Internet search of the full action name. For example, cas.builtins.addNode, and the first result usually leads you to the action document.

In addition to the documentation, you can also use tab completion to display what is in an action set.

```
> cas.builtins.<tab>
```

```
cas.builtins.about              cas.builtins.actionSetInfo
cas.builtins.addNode            cas.builtins.casCommon
cas.builtins.echo               cas.builtins.getLicensedProductInfo
cas.builtins.getLicenseInfo     cas.builtins.help
cas.builtins.history            cas.builtins.httpAddress
cas.builtins.installActionSet   cas.builtins.listNodes
cas.builtins.loadActionSet      cas.builtins.log
cas.builtins.modifyQueue        cas.builtins.ping
cas.builtins.queryActionSet     cas.builtins.queryName
cas.builtins.reflect            cas.builtins.refreshLicense
[...truncated]
```

Now that we have seen how to query the server for available action sets and actions, and we know how to get help for the actions, we can move on to some more advanced action calls.

Specifying Action Parameters

We have already seen a few action parameters being used on the builtins.help action (action, actionset, and verbose). Now, let's look at the descriptions of the parameters.

```
> out <- cas.builtins.help(conn, action = 'help')
```

```
NOTE: Information for action 'builtins.help':
NOTE: The following parameters are accepted.  Default values are shown.
NOTE:     string action=NULL,
NOTE:        specifies the name of the action for which you want help. The name can
be in the form 'actionSetName.actionName' or just 'actionName.
NOTE:     string actionSet=NULL,
NOTE:        specifies the name of the action set for which you want help. This
parameter is ignored if the action parameter is specified.
NOTE:     boolean verbose=true
NOTE:        when set to True, provides more detail for each parameter.
```

You see that action and actionset are declared as strings, and verbose is declared as a Boolean. Action parameters can take many types of values. Table 3.1 shows the supported types:

Table 3.1: Supported Types of Parameters

CAS Type	R Type	Description
Boolean	logical	Value that indicates true or false. This should always be specified using R's TRUE or FALSE values.
double	numeric	64-bit floating-point number
int32	integer	32-bit integer
int64	integer	64-bit integer
string	character	Character content. Note that if a byte string is passed as an argument, SWAT attempts to convert it to Unicode using the default encoding.
value list	vector or list	Collection of items. R vectors become indexed CAS value lists. R lists become keyed CAS value lists.

The easiest way to practice more complex arguments is by using the builtins.echo action. This action simply prints the values of all parameters that were specified in the action call. The following code demonstrates the builtins.echo action with all of the parameter types in Table 3.1.

```
> out <- cas.builtins.echo(
+    conn,
+    boolean_true = TRUE,
+    boolean_false = FALSE,
+    double = 3.14159,
+    int32 = 1776,
+    int64 = 2**60,
+    string = 'I like snowmen! \u2603',
+    vector = c('item1', 'item2', 'item3'),
+    list = list(key1 = 'value1',
+                key2 = 'value2',
+                key3 = 3)
+ )
```

```
NOTE: builtin.echo called with 9 parameters.
NOTE:     parameter 1: _messagelevel = 'note'
NOTE:     parameter 2: boolean_false = false
NOTE:     parameter 3: boolean_true = true
NOTE:     parameter 4: double = 3.1416
NOTE:     parameter 5: int32 = 1776
NOTE:     parameter 6: int64 = 1.15292e+18
NOTE:     parameter 7: list = {key1 = 'value1', key2 = 'value2', key3 = 3}
NOTE:     parameter 8: string = 'I like snowmen! <U+2603>'
NOTE:     parameter 9: vector = {'item1', 'item2', 'item3'}
```

You might notice that the parameters are printed in a different order than what was specified in the echo call. This is simply because keyword parameters in R are stored in a list, and lists don't keep keys in a specified order.

You might also notice that the printed syntax is not R syntax. It is a pseudo-code syntax that is more similar to the Lua programming language. Lua is used in other parts of CAS as well (such as the builtins.history action), so most code-like objects that are printed from CAS are in Lua or syntax that is like Lua. However, the syntax of the two languages (as far as parameter data structures goes) are similar enough that it is easy to see the mapping from one to the other. The biggest differences are in the value list parameters. Vectors in the printout use braces, whereas R uses the c function. Also, in the keyed list, R uses syntax list() rather than braces.

The complexity of the parameter structures is unlimited. Vectors can be nested inside lists, and lists can be nested inside vectors. A demonstration of nested structures in echo follows:

```
> out <- cas.builtins.echo(
+    conn,
+    list = list(
+      'item1',
+      'item2',
+      list(
+        key1 = 'value1',
+        key2 = list(
+          value2 = c(0, 1, 1, 2, 3)
+          )
```

```
+          )
+          ))
```

```
NOTE: builtin.echo called with 2 parameters.
NOTE:     parameter 1: _messagelevel = 'note'
NOTE:     parameter 2: list = {'item1', 'item2', {key1 = 'value1', key2 = {value2 =
{0, 1, 1, 2, 3}}}}
```

There are a couple of features of the CAS parameter processor that can make your life a bit easier. We look at those in the next section.

Automatic Type Casting

So far, we have constructed arguments using either the exact data types expected by the action or the arbitrary parameters in echo. However, the CAS action parameter processor on the server is flexible enough to enable passing in parameters of various types. If possible, those parameters are converted to the proper type before they are used by the action.

The easiest form of type casting to demonstrate is the conversion of strings to numeric values. If an action parameter takes a numeric value, but you pass in a string that contains a numeric representation as its content, the CAS action processor parses out the numeric value and sends that value to the action. This behavior can be seen in the following action calls to builtins.history, which shows the action call history. The first call uses integers for first and last, but the second call uses strings. In either case, the result is the same due to the automatic conversion on the server side.

```
# Using integers
> out <- cas.builtins.history(conn, first = 5, last = 7)
```

```
NOTE: 5: action builtins.actionSetInfo / extensions={'accessControl'},
_messageLevel='error'; /* (SUCCESS) */
NOTE: 6: action builtins.listActions / actionSet='accessControl',
_messageLevel='error'; /* (SUCCESS) */
NOTE: 7: action builtins.actionSetInfo / extensions={'builtins'},
_messageLevel='error'; /* (SUCCESS) */
```

```
# Using strings as integer values
> out <- cas.builtins.history(conn, first = '5', last = '7')
```

```
NOTE: 5: action builtins.actionSetInfo / extensions={'accessControl'},
_messageLevel='error'; /* (SUCCESS) */
NOTE: 6: action builtins.listActions / actionSet='accessControl',
_messageLevel='error'; /* (SUCCESS) */
NOTE: 7: action builtins.actionSetInfo / extensions={'builtins'},
_messageLevel='error'; /* (SUCCESS) */
```

Although the server can do some conversions between types, it is generally a good idea to use the correct type. There is another type of automatic conversion that adds syntactical enhancement to action calls. This is the conversion of a scalar-valued parameter to a list value. This is described in the next section.

Scalar Parameter to List Conversion

Many times, when using an action parameter that requires a list as an argument, you use only the first key in the list to specify the parameter. For example, the builtins.history action takes a parameter called casOut.

This parameter specifies an output table to put the history information into. The specification for this parameter follows:

```
> out <- cas.builtins.help(conn, action = 'history')
```

```
... truncated ...

NOTE:       list casOut={
NOTE:           specifies the settings for saving the action history to an
                output table.
NOTE:           string name=NULL,
NOTE:           specifies the name to associate with the table.
NOTE:           string caslib=NULL,
NOTE:           specifies the name of the caslib to use.
NOTE:           string timeStamp=NULL,
NOTE:           specifies the timestamp to apply to the table. Specify the
                value in the form that is appropriate for your session locale.
NOTE:           boolean compress=false,
NOTE:           when set to True, data compression is applied to the table.
NOTE:           boolean replace=false,
NOTE:           specifies whether to overwrite an existing table with the same name.

... truncated ...
```

The first key in the casOut parameter is name and indicates the name of the CAS table to create. The complete way of specifying this parameter with only the name key follows:

```
> out <- cas.builtins.history(conn, casout = list(name='hist'))
```

This is such a common idiom that the server enables you to specify list values with only the first specified key given (for example, name), just using the value of that key. That is a mouthful, but it is easier than it sounds. It just means that rather than having to use the list to create a nested list, you could simply do the following:

```
> out <- cas.builtins.history(conn, casout = 'hist')
```

Of course, if you need to use any other keys in the casOut parameter, you must use the list form. This conversion of a scalar value to a list value is common when specifying input tables and variable lists of tables, which we see later on.

Now that we have spent some time on the input side of CAS actions, let's look at the output side.

CAS Action Results

Up to now, all of our examples have stored the result of the action calls in a variable, but we have not done anything with the results yet. Let's start by using our example of all of the CAS parameter types.

```
> out <- cas.builtins.echo(
+    conn,
+    boolean_true = TRUE,
+    boolean_false = FALSE,
+    double = 3.14159,
+    int32 = 1776,
+    int64 = 2**60,
```

```
+     string = 'I like snowmen! \u2603',
+     vector = c('item1', 'item2', 'item3'),
+     list = list(key1 = 'value1',
+                  key2 = 'value2',
+                  key3 = 3)
+ )
```

```
NOTE: builtin.echo called with 9 parameters.
NOTE:     parameter 1: _messagelevel = 'note'
NOTE:     parameter 2: boolean_false = false
NOTE:     parameter 3: boolean_true = true
NOTE:     parameter 4: double = 3.1416
NOTE:     parameter 5: int32 = 1776
NOTE:     parameter 6: int64 = 1.15292e+18
NOTE:     parameter 7: list = {key1 = 'value1', key2 = 'value2', key3 = 3}
NOTE:     parameter 8: string = 'I like snowmen! <U+2603>'
NOTE:     parameter 9: vector = {'item1', 'item2', 'item3'}
```

Displaying the contents of the out variable gives the following:

```
> out
```

```
$`_messagelevel`
[1] "note"

$boolean_false
[1] FALSE

$boolean_true
[1] TRUE

$double
[1] 3.1416

$int32
[1] 1776

$int64
[1] 1.152922e+18

$list
$list$key1
[1] "value1"

$list$key2
[1] "value2"

$list$key3
[1] 3


$string
[1] "I like snowmen! ☃"

$vector
$vector[[1]]
[1] "item1"
```

```
$vector[[2]]
[1] "item2"

$vector[[3]]
[1] "item3"
```

```
> class(out)
```

```
[1] "list"
```

The object that is held in the out variable is a list object. You can traverse and modify the result just as you could any other R list object. For example, if you wanted to walk through the items and print each key and value explicitly, you could do the following:

```
> for (key in names(out)){
+    print(key)
+    print(out[[key]])
+    cat('\n')
+ }
```

```
[1] "_messagelevel"
[1] "note"

[1] "boolean_false"
[1] FALSE

[1] "boolean_true"
[1] TRUE

[1] "double"
[1] 3.1416

[1] "int32"
[1] 1776

[1] "int64"
[1] 1.152922e+18

[1] "list"
$key1
[1] "value1"

$key2
[1] "value2"

$key3
[1] 3


[1] "string"
[1] "I like snowmen! ☃"

[1] "vector"
[[1]]
[1] "item1"
```

```
[[2]]
[1] "item2"

[[3]]
[1] "item3"
```

Although the object that is returned by an action is always a list object, the contents of that object depend completely on the purpose of that action. It could be as simple as key/value pairs of scalars and as complex as a nested structure of lists, such as our parameters in the previous section. Actions that perform analytics typically return one or more data.frames that contain the results.

Since the results objects are simply R lists, we assume that you are able to handle operations on them. But we will take a closer look at the casDataFrame in the next section.

Using casDataFrames

The casDataFrames that are returned by CAS actions are extensions of the data.frames that are defined by R. Largely, both work the same way. The only difference is that the casDataFrame returned by CAS contains extra metadata that is found in typical SAS data sets. This metadata includes things such as SAS data format names, the SAS data type, and column and table labels.

One of the builtins actions that returns a casDataFrame is builtins.help. This action returns a casDataFrame that is filled with the names and descriptions of all the actions that are installed on the server. Each action set gets its own key in the result. Let's look at some output from builtins.help.

The following code runs the builtins.help action, lists the names in the list object that is returned, verifies that it is a casDataFrame object using R's Class function, and displays the contents of the casDataFrame (some output is reformatted slightly for readability):

```
> out <- cas.builtins.help(conn)

> names(out)
 [1] "accessControl"  "builtins"       "configuration"  "dataPreprocess"
 [5] "dataStep"       "percentile"     "search"         "session"
 [9] "sessionProp"    "simple"         "table"

> class(out[['builtins']])
[1] "casDataFrame"
attr(,"package")
[1] "swat"

> out[['builtins']]
```

	name	description
1	addNode	Adds a machine to the server
2	removeNode	Remove one or more machines from the...
3	help	Shows the parameters for an action o...
4	listNodes	Shows the host names used by the server
5	loadActionSet	Loads an action set for use in this ...
6	installActionSet	Loads an action set in new sessions ...
7	log	Shows and modifies logging levels
8	queryActionSet	Shows whether an action set is loaded
9	queryName	Checks whether a name is an action o...
10	reflect	Shows detailed parameter information...

```
11        serverStatus              Shows the status of the server
12               about              Shows the status of the server
13            shutdown                      Shuts down the server
14            userInfo  Shows the user information for your ...
15       actionSetInfo  Shows the build information from loa...
16             history  Shows the actions that were run in t...
17           casCommon  Provides parameters that are common ...
18                ping  Sends a single request to the server...
19                echo  Prints the supplied parameters to th...
20         modifyQueue  Modifies the action response queue s...
21      getLicenseInfo  Shows the license information for a ...
22      refreshLicense  Refresh SAS license information from...
23         httpAddress  Shows the HTTP address for the serve...
```

casDataFrame Operations

We can store this casDataFrame in another variable to make it a bit easier to work with. Because the returned results are list objects, you can access keys as attributes. This means that we can access the builtins key of the out variable in either of the following ways:

```
> blt <- out[['builtins']]

> blt <- out$builtins
```

Which syntax you use depends on personal preference. The $ syntax is a bit cleaner, but the bracketed syntax works regardless of the key value (including white space, or other special characters). Typically, you might use the $ syntax in interactive programming, but the bracketed syntax is better for production code.

Now that we have a handle on the casDataFrame, we can do typical data.frame operations on it, such as sorting and filtering. For example, to sort the builtins actions by the name column, you might do the following.

```
> blt[order(blt$name),]
```

```
              name                      description
12           about              Shows the status of the server
15   actionSetInfo  Shows the build information from loa...
1          addNode                 Adds a machine to the server
17       casCommon  Provides parameters that are common ...
19            echo  Prints the supplied parameters to th...
21  getLicenseInfo  Shows the license information for a ...
3             help  Shows the parameters for an action o...
16         history  Shows the actions that were run in t...
23     httpAddress  Shows the HTTP address for the serve...
6   installActionSet  Loads an action set in new sessions ...
4        listNodes  Shows the host names used by the server
5    loadActionSet  Loads an action set for use in this ...
7              log        Shows and modifies logging levels
20     modifyQueue  Modifies the action response queue s...
18            ping  Sends a single request to the server...
8    queryActionSet     Shows whether an action set is loaded
9        queryName  Checks whether a name is an action o...
10         reflect  Shows detailed parameter information...
22  refreshLicense  Refresh SAS license information from...
2       removeNode  Remove one or more machines from the...
```

```
11        serverStatus           Shows the status of the server
13        shutdown                     Shuts down the server
14        userInfo     Shows the user information for your ...
```

As you can see, working with results from CAS actions is the same as the workflow with any other R framework. You connect to a CAS host, run a CAS action, and process the list object that is returned.

Now that we understand the basics of the workflow, let's look at how to add additional action sets and actions to your CAS session.

Working with CAS Action Sets

In the previous sections, we have already seen that a CAS session has access to multiple action sets that each contain multiple actions. However, all of the action sets that we have seen so far have been loaded automatically when we connect to CAS. We haven't shown how to load additional action sets in order to do additional operations, such as advanced analytics, machine learning, streaming data analysis, and so on.

In order to load new action sets, we must first see what action sets are available on our server. We can use the builtins.actionsetinfo action to do that. We are going to use the all=TRUE option to see all of the action sets that are installed on the server, rather than only the ones that are currently loaded.

```
# Run the builtins.actionsetinfo action.
> asinfo <- cas.builtins.actionSetInfo(conn, all = TRUE)

# Filter the casDataFrame to contain only action sets that
# have not been loaded yet.
> asinfoNotLoaded <- asinfo$setinfo[asinfo$setinfo$loaded == 0, ]

# Create a new casDataFrame with only columns between
# actionset and label.
> asinfoNotLoaded <- asinfoNotLoaded[,c('actionset','label')]

> asinfoNotLoaded
```

	actionset	label
1	access	
3	actionTest	
4	actionTest2	
5	aggregation	
6	astore	
7	autotune	
8	bayesianNetClassifier	
...	...	...
120	testsscp	Actions for SSCP Computation
121	riskmethods	riskMethods
122	dmcastest	Data Management Test Services
123	risksim	Risk Simulation
124	clarify	Call Tool API routines
125	csptest	CASCLP
126	risk	RiskResults
127	cmpcas	
128	sandcas	SANDWICH library
129	tkcasls	CASL Server
130	riskut	RiskUT

Depending on your installation and licensing, the list varies from system to system. One very useful action set that should be automatically available on all systems is the simple action set. This action set contains actions for simple statistics, such as summary statistics (max, min, mean, and so on), correlations, and frequencies. To load an action set, use the builtins.loadActionSet action:

```
> cas.builtins.loadActionSet(conn, 'simple')
```

```
NOTE: Added action set 'simple'.
NOTE: Information for action set 'simple':
NOTE:    simple
NOTE:        mdSummary - Calculates multidimensional summaries of numeric variables
NOTE:        numRows - Shows the number of rows in a Cloud Analytic Services table
NOTE:        summary - Generates descriptive statistics of numeric variables such as
the sample mean, sample variance, sample size, sum of squares, and so on
NOTE:        correlation - Computes Pearson product-moment correlations.
NOTE:        regression - Performs a linear regression up to 3rd-order polynomials
NOTE:        crossTab - Performs one-way or two-way tabulations
NOTE:        distinct - Computes the distinct number of values of the variables in
the variable list
NOTE:        topK - Returns the top-K and bottom-K distinct values of each variable
included in the variable list based on a user-specified ranking order
NOTE:        groupBy - Builds BY groups in terms of the variable value combinations
given the variables in the variable list
NOTE:        freq - Generates a frequency distribution for one or more variables
NOTE:        paraCoord - Generates a parallel coordinates plot of the variables in
the variable list

$actionset
[1] "simple"
```

As you can see, this action returns a list object as described in the previous section. It contains a single key called actionset that contains the name of the action set that was loaded. Typically, you do not need this return value, but it can be used to verify that the action set has been loaded. If you attempt to load an action set that cannot be loaded for some reason (such as incorrect name, no license, or no authorization), the list object is empty.

Now that we have loaded the simple action set, we can get help on it using the usual builtins.help action or refer to the SAS Help Center.

```
> out <- cas.builtins.help(conn, actionset = 'simple')
NOTE: Information for action set 'simple':
NOTE:    simple
NOTE:        mdSummary - Calculates multidimensional summaries of numeric
variables
NOTE:        numRows - Shows the number of rows in a Cloud Analytic Services table
NOTE:        summary - Generates descriptive statistics of numeric variables such
as the sample mean, sample variance, sample size, sum of squares, and so on
NOTE:        correlation - Computes Pearson product-moment correlations.
NOTE:        regression - Performs a linear regression up to 3rd-order polynomials
NOTE:        crossTab - Performs one-way or two-way tabulations
NOTE:        distinct - Computes the distinct number of values of the variables in
the variable list
NOTE:        topK - Returns the top-K and bottom-K distinct values of each
variable included in the variable list based on a user-specified ranking order
NOTE:        groupBy - Builds BY groups in terms of the variable value
combinations given the variables in the variable list
```

```
NOTE:       freq - Generates a frequency distribution for one or more variables
NOTE:       paraCoord - Generates a parallel coordinates plot of the variables in
the variable list
```

To get to the SAS Help Center, you can use the help function or the ? suffix operator on any CAS action, and a "Common Page for CAS Actions" will be displayed. In that common page, there are links to documents of CAS actions by name, action sets by name, and action sets by product. For example, you can use the help function on the simple.correlation action to open the common help page.

```
> help(cas.simple.correlation)
```

Or you can just use the ? suffix operator.

```
> ?cas.simple.correlation
```

Once an action set has been loaded, it cannot be unloaded. The overhead for keeping an action set loaded is minimal, so this issue doesn't make a significant difference.

That is really all there is to loading action sets. We still do not have data in our system, so we cannot use any of the simple statistics actions yet. Let's review some final details about options and dealing with errors in the next section. Then, the following chapter gets into the ways of loading data and using the analytical actions on those data sets.

Details

We have covered the overall workings of connecting to a CAS host, running CAS actions, working with the results of CAS actions, and loading CAS action sets. However, there are some details that we haven't covered. Although these items aren't necessary for using SWAT and CAS, they can be quite useful to have in your tool belt.

Getting Help

Even though we have already covered the methods for getting help from CAS, it is an important topic to recap. Every object in the SWAT package uses the standard R method of surfacing documentation. This includes the help function in R (for example, help(CAS)), the ? suffix operator (for example, ?CAS), and any other tool that uses R's docstrings.

Due to their run-time nature, action sets and actions do not support the same R help function and ? suffix. However, you can use the builtins.help and the SAS Help Center online.

Keep in mind that tab completion on the CAS objects, CAS actions, and other objects in the SWAT package can be a quick reminder of the attributes, parameters, and methods of that object.

These help system hooks should be sufficient to help you get information about any objects in the SWAT package, CAS action sets, and CAS actions and their parameters. If more detailed information is needed, it is available in the official SAS Viya documentation in the SAS Help Center.

CAS Session Options

The session on the CAS host has options that can be set to change certain behaviors for the current session. These options are set using the sessionProp.setSessOpt action. You can view them and get current values using sessionProp.listSessOpts and sessionProp.getSessOpt. The best way to see all of the options that are available is to check the action syntax in SAS Help Center. For example, the following is a list of options that are available to the sessionProp.setSessOpt action:

```
appTag="string"
```
        specifies the string to prefix to log messages.

```
caslib="string"
```
        specifies the caslib name to set as the active caslib.

```
cmpopt="string"
```
        specifies code generation optimizations.

```
collate="MVA" | "UCA"
```
        specifies the collating sequence for sorting.

        Default   UCA
        MVA
                specifies SAS client collating.

        UCA
                specifies a locale-appropriate collating sequence.

```
dataStepFmtErr=TRUE | FALSE
```
        when set to True, the DATA step generates an error when a variable
        format cannot be found. You can set this parameter to False to continue
        processing instead of generating an error.

        Alias   fmtErr
        Default   TRUE
```
dataStepMsgSumLevel="ALL" | "NONE" | "PUT"
```
        specifies the DATA step message summary level. When the DATA step runs
        on multiple threads, the same message can be generated on each thread.
        This parameter controls the summary level of duplicate messages.

        Default   ALL
        ALL
                specifies that the first occurrence of all messages and PUT
                statements are sent to the client when they occur. Duplicate
                messages and PUT statements are summarized and sent to the
                client when the DATA step exits. This is the default.

        NONE
                specifies that all message and PUT statements, from every
                thread, are sent to the client log without any summarization.

        PUT
                specifies that the first occurrence of all message and PUT
                statements are sent to the client. Duplicate messages are
                summarized and sent to the client when the DATA step exits. PUT
                statements are not summarized. They are sent to the client when
                they occur.

```
dataStepReplaceTable=TRUE | FALSE
```
        specifies whether DATA step can replace an existing table.

        Default   TRUE

dqLocale="string"
>>   specifies the default locale to use for data quality operations. Use a
        5-letter locale code.

dqQKBRepoLoc="string"
>   specifies the name of the caslib where QKBs are stored.

>   Default   "ReferenceData"
dqSetupLoc="string"
>>   specifies the name of the default QKB to use for data quality
        operations.

eventds="string"
>   specifies one or more event objects that define custom date events.

intervalds="string"
>   specifies interval object pairs.

locale="string"
>   specifies the locale to use for sorting and formatting.

>   Default   "en_US"
logFlushTime=64-bit-integer
>>   specifies the log flush time, in milliseconds. A value of -1 indicates
        to flush logs after each action completes. A value of 0 indicates to
        flush logs as they are produced.

>   Default   100
>   Range   -1-86400
>>   TIP   You can enclose the value in quotation marks and specify B, K,
        M, G, or T as a suffix to indicate the units. For example, "8M"
        specifies eight megabytes.
maxTableMem=64-bit-integer
>>   specifies the maximum amount of memory, in bytes, that each thread
        should allocate for in-memory blocks before converting to a memory-
        mapped file. Files are written in the directories that are specified in
        the CAS_DISK_CACHE environment variable.

>   Default   16777216
>>   TIP   You can enclose the value in quotation marks and specify B, K,
        M, G, or T as a suffix to indicate the units. For example, "8M"
        specifies eight megabytes.
messageLevel="ALL" | "DEFAULT" | "ERROR" | "NONE" | "NOTE" | "WARNING"
>   specifies the log message level.

>   Default   ALL
metrics=TRUE | FALSE
>   when set to True, action metrics are displayed.

>   Default   FALSE

... truncated ...

There are options for setting the session locale, collation order, time-outs, memory limits, and so on. The metrics option is simple to demonstrate. Let's get its current value using sessionProp.getSessOpt:

```
> out <- cas.sessionProp.getSessOpt(conn, name = 'metrics')
> out
```

```
$metrics
[1]  0
```

The output is our usual list object with a key that matches the requested option name. In this case, the metrics option is returned as an integer value of zero (corresponding to a Boolean false). You can get the actual value of the metrics option by accessing that key from the list object.

```
> out$metrics
```

```
[1]  0
```

Setting the values of options is done using sessionProp.setSessOpt with keyword arguments for the option names. You can specify as many options in sessionProp.setSessOpt as you need.

```
> cas.sessionProp.setSessOpt(conn, metrics = TRUE, collate = 'MVA')
NOTE: Executing action 'sessionProp.setSessOpt'.
NOTE: Action 'sessionprop.setsessopt' used (Total process time):
NOTE:         real time            0.000370 seconds
NOTE:         cpu time             0.000000 seconds (0.00%)
NOTE:         total nodes          1 (32 cores)
NOTE:         total memory         188.99G
NOTE:         memory               98.19K (0.00%)
```

Notice that the metrics option takes effect immediately. We now get performance metrics of the action that are printed to the output. Checking the value of collate, you see that it has been set to MVA.

```
> out <- cas.sessionProp.getSessOpt(conn, name = 'collate')
NOTE: Executing action 'sessionProp.getSessOpt'.
NOTE: Action 'sessionprop.getsessopt' used (Total process time):
NOTE:         real time            0.000302 seconds
NOTE:         cpu time             0.000000 seconds (0.00%)
NOTE:         total nodes          1 (32 cores)
NOTE:         total memory         188.99G
NOTE:         memory               49.91K (0.00%)
```

```
> out$collate
```

```
[1]  "MVA"
```

Conclusion

We have covered a lot of territory in this chapter, but you should now have the tools that you need in order to connect to CAS, call CAS actions, and traverse the results. Finally, we demonstrated some of the SWAT client and CAS session options to control certain behaviors of both areas. Now that we have that all out of the way, we can move on to something a little more interesting: data and how to get it into CAS.

---

[1] Technically, these parameters can also be specified by setting environment variables CASHOST and CASPORT, and not specified in the CAS constructor.

Chapter 4: Managing Your Data in CAS

Before you can do any sort of analysis in CAS, you need some data to work with. There are two components to data access in CAS: caslibs and CAS tables. Caslibs are definitions that give access to a resource that contains data. These resources can be files that are located in a file system, a database, streaming data from an ESP (Event Stream Processing) server, or other data sources that SAS can access. Caslibs contain all of the connection information, as well as authentication and authorization settings. In addition, caslibs are containers for tables that are loaded in the server, effectively enabling you to create namespaces for tables.

When you want to analyze data from one of your caslib resources, you load the data into a CAS table. A CAS table contains columns of data and information about the data in the columns (including column labels, data types, data formats, and so on).

CAS tables and caslibs are the topic of this chapter.

Overview

The subject of caslibs and CAS tables is fairly extensive, especially when you consider the multiple ways of loading and parsing data from various sources. We have attempted to boil it down to one simple diagram (Figure 4.1) to give you an overall view of the process and components that are involved.

Figure 4.1: Data Processing Flow Diagram

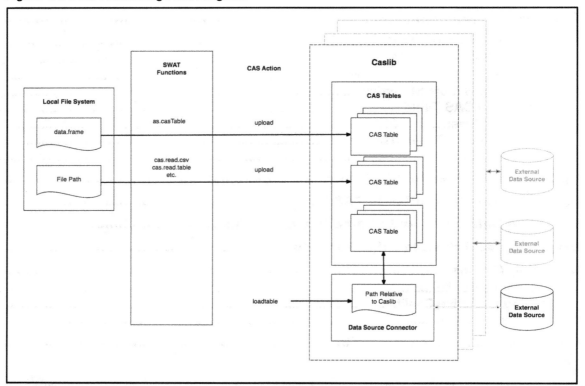

In Figure 4.1, you can see that SWAT has multiple methods for loading data, and the CAS actions that are used to load the data vary as well. Each session can have one or more caslibs, with each connected to a different data source. The caslibs and the tables within the caslibs might have session scope or global scope, depending on the options that they were created with. These details are covered in the following sections.

Getting Started with Caslibs and CAS Tables

Since all data that is accessed by CAS must be in a caslib, it makes sense to start there when discussing how to get your data into CAS. Depending on your CAS server setup, you might already have one or more caslibs configured and ready to use. To find out what caslibs are available, use the table.caslibInfo action.

```
> library('swat')
> conn <- CAS('server-name.mycompany.com', port = port-number,
              username = 'userid', password = 'password')
> cas.table.caslibInfo(conn)
```

```
$CASLibInfo
            Name    Type                   Description
1 CASUSER(username)  PATH Personal File System Caslib
2          Formats  PATH                Format Caslib
                  Path    Definition Subdirs Local Active
1          /u/username/                    1      0      1
2    /bigdisk/lax/formats/                 1      0      0
   Personal Hidden Transient
1         1      0         1
2         0      0         0
```

We can see from this that we have two caslibs defined already. The first caslib points to our home directory. The second caslib is for holding user-defined data formats. Both of these initial caslibs are set up by a system administrator. Depending on your permissions, you might be able to add others as needed. You see that the table describing the caslibs also contains additional information about each caslib.

A human-readable description can be added when creating a caslib, as well as information about the breadth and scope of the caslib. The Subdirs column indicates whether subdirectories of the specified path of the caslib can be accessed. A value of 0 means no, and a value of 1 means yes. The Local column indicates whether the caslib is visible in the current session (indicated by 1), or if it is visible by all sessions (indicated by 0). The Active column indicates which caslib is active. When a caslib is active, it is used if a caslib is not explicitly specified during any operation that requires one. The Personal column indicates whether the caslib is specific to the user. These caslibs can be seen only by the current user regardless of whether the caslib is local or global in scope. Finally, the Hidden column indicates whether the caslib should be displayed in caslib listings. The caslib is displayed only when the showhidden parameter is set to TRUE in the caslibinfo action.

Now that we know what caslibs are available to us, let's see what they contain. We have some data files that are stored in the data subdirectory of our home directory, which is pointed to by the caslib Casuser. Let's use the table.fileInfo action to list the contents of that directory.

```
> cas.table.fileInfo(conn, caslib = 'casuser', path = 'data')
```

```
$FileInfo
     Permission    Owner Group                        Name
   1 -rw-r--r--  username users                    iris.csv
   2 -rw-r--r--  username users                    cars.csv
   3 -rw-r--r--  username users     sashelp_class.sashdat
   4 -rw-r--r--  username users                   class.csv
```

```
        Size Encryption                    Time
1       3716                 18Feb2016:17:25:43
2      42177                 18Feb2016:17:25:49
3      82136        NONE     19Feb2016:17:02:37
4        519                 14Apr2016:10:39:02
```

You'll notice that when we specified the caslib to look in, we used Casuser, but we didn't include the username enclosed in parentheses. The username part can be specified, but it is primarily there to differentiate multiple Casuser caslibs in the listing. You'll also notice that caslib names, as well as CAS table names, are not case sensitive. In addition, since Casuser is our active caslib, we don't have to specify a caslib option at all.

```
> cas.table.fileInfo(conn, path = 'data')
```

```
$FileInfo
    Permission       Owner   Group                         Name
1   -rw-r--r--    username   users                     iris.csv
2   -rw-r--r--    username   users                     cars.csv
3   -rw-r--r--    username   users     sashelp_class.sashdat
4   -rw-r--r--    username   users                    class.csv

        Size Encryption                    Time
1       3716                 18Feb2016:17:25:43
2      42177                 18Feb2016:17:25:49
3      82136        NONE     19Feb2016:17:02:37
4        519                 14Apr2016:10:39:02
```

Although you can omit the caslib option, it's generally a good practice to include it. When you create larger programs and start moving code around, your active caslib might not be what was in your original program context.

In addition to the files that are accessible to this caslib, there can also be tables that have already been loaded that are available to the caslib. These can be seen using the table.tableInfo action.

```
> out <- cas.table.tableInfo(conn)
```

```
NOTE: No tables are available in caslib CASUSER(username) of Cloud Analytic
Services.
```

At this point, we don't have any tables that are already loaded. Since we have data files that are located in an area that is accessible to our Casuser caslib, let's load one into a table.

Loading Data into a CAS Table

To load data from a file in a caslib into a CAS table, you use the table.loadTable action by specifying the file path and the caslib name.

```
> out <- cas.table.loadTable(
    conn,
    path = 'data/iris.csv',
    caslib='casuser'
    )
```

```
NOTE: Cloud Analytic Services made the file data/iris.csv available as table
DATA.IRIS in caslib CASUSER(username).
```

```
> out
```

```
$caslib
[1] "CASUSER(username)"

$tableName
[1] "DATA.IRIS"
```

You might remember from Chapter 2 that we used the as.casTable function to load data. That function loads data from the client side, whereas the table.loadTable action loads files from the server side. We'll get into more details about the as.casTable function later in this chapter.

The result of the table.loadTable action includes the table name (tableName) and the caslib (caslib). The table name that you get from the table.loadTable action is generated and is based on the file name that was loaded. You can change the name of the loaded table using the casout option. In the following example, we specify both the name and the caslib that the loaded table is written to:

```
>out <- cas.table.loadTable(
   conn,
   path = 'data/iris.csv',
   caslib='casuser',
   casout=list(name='mydata', caslib='casuser')
   )
```

```
NOTE: Cloud Analytic Services made the file data/iris.csv available as table MYDATA
in caslib CASUSER(username).
```

```
> out
```

```
$caslib
[1] "CASUSER(username)"

$tableName
[1] "MYDATA"
```

Now that our data is loaded into a table, let's request information about the table from the server. The first action you'll want to use is the table.tableInfo action.

```
> cas.table.tableInfo(conn, name = 'data.iris', caslib = 'casuser')
```

```
$TableInfo

        Name   Rows   Columns  Encoding CreateTimeFormatted  \
 0   DATA.IRIS   150         5     utf-8  02Nov2016:11:54:28

     ModTimeFormatted JavaCharSet    CreateTime         ModTime  \
 0   02Nov2016:11:54:28        UTF8  1.793707e+09  1.793707e+09

    Global  Repeated  View    SourceName         SourceCaslib  \
 0       0         0     0  data/iris.csv  CASUSER(username)
```

```
       Compressed   Creator Modifier
   0            0   username
```

From these results, we see that there are 150 rows of data in 5 columns. Another field of particular note is the Global value field, which indicates whether the table can be seen in other CAS sessions. In our case, the value is 0, and therefore, it is local to this session only.

Now that we have some information about the table as a whole, let's see what information we can get about the columns. To get information about CAS table columns, you use the table.columnInfo action. Since the table.tableInfo action gives a broad view of the table information, it needs only a table name and a caslib name. The table.columnInfo action is somewhat more complex and uses a dictionary as a table argument.

```
> out <- cas.table.columnInfo(
    conn,
    table = list(name = 'data.iris', caslib = 'casuser')
    )
> out
```

```
$ColumnInfo

          Column  ID     Type  RawLength  FormattedLength  NFL  NFD
0   Sepal.Length   1   double          8               12    0    0
1    Sepal.Width   2   double          8               12    0    0
2   Petal.Length   3   double          8               12    0    0
3    Petal.Width   4   double          8               12    0    0
4        Species   5  varchar         10               10    0    0
```

From this output, you can see the column names, the data types, the data sizes, and various formatting information.

Now that we know about the metadata of our data, let's look at the data itself.

Displaying Data in a CAS Table

Of course, the most important thing about the table is the data itself. We can get a sample of the data by using the table.fetch action. In this case, we use the to=5 parameter to fetch only five rows.

```
> cas.table.fetch(
    conn,
    table = list(name = 'data.iris', caslib = 'casuser'),
    to = 5
    )
```

```
$Fetch
  _Index_  Sepal.Length  Sepal.Width  Petal.Length  Petal.Width        Species
1       1           5.1          3.5           1.4          0.2    Iris-setosa
2       2           4.9          3.0           1.4          0.2    Iris-setosa
3       3           6.7          3.3           5.7          2.5 Iris-virginica
4       4           4.7          3.2           1.3          0.2    Iris-setosa
5       5           4.6          3.1           1.5          0.2    Iris-setosa
```

Note that unless you specify sorting options, the order of the data that comes back is not guaranteed to be in the same order that the data was entered in. This is because data can be spread across the CAS grid and the result that you see is the compiled version of all of the results of the nodes on a grid. In the following code, we fetch the data using the Sepal.Length and Sepal.Width sort keys to guarantee ordering.

```
> cas.table.fetch(
    conn,
    table = list(name = 'data.iris', caslib = 'casuser'),
    to = 5,
    sortby=list('Sepal.Length ', 'Sepal.Width')
    )
```

```
$Fetch
 _Index_ Sepal.Length Sepal.Width Petal.Length Petal.Width    Species
1      1          4.3         3.0          1.1         0.1 Iris-setosa
2      2          4.4         2.9          1.4         0.2 Iris-setosa
3      3          4.4         3.0          1.3         0.2 Iris-setosa
4      4          4.4         3.2          1.3         0.2 Iris-setosa
5      5          4.5         2.3          1.3         0.3 Iris-setosa
```

Now that we have seen a sample of the data, let's compute some simple statistics.

Computing Simple Statistics

Just to whet your appetite a bit, let's get some summary statistics on the table that we just loaded. We'll delve into running actions on data in later chapters, but it seems a shame to load our first set of data and not do any sort of analysis on it. Let's run the simple.summary action on our table.

```
> cas.simple.summary(
    conn,
    table = list(name='data.iris', caslib='casuser')
    )
```

```
$Summary
        Column Min Max   N NMiss     Mean   Sum      Std
1 Sepal.Length 4.3 7.9 150     0 5.843333 876.5 0.8280661
2  Sepal.Width 2.0 4.4 150     0 3.054000 458.1 0.4335943
3 Petal.Length 1.0 6.9 150     0 3.758667 563.8 1.7644204
4  Petal.Width 0.1 2.5 150     0 1.198667 179.8 0.7631607

       StdErr       Var     USS       CSS       CV    TValue
1 0.06761132 0.6856935 5223.85 102.16833 14.17113 86.42537
2 0.03540283 0.1880040 1427.05  28.01260 14.19759 86.26430
3 0.14406432 3.1131794 2583.00 463.86373 46.94272 26.09020
4 0.06231181 0.5824143  302.30  86.77973 63.66747 19.23659

         ProbT  Skewness   Kurtosis
1 3.331256e-129  0.3149110 -0.5520640
2 4.374977e-129  0.3340527  0.2907811
3  1.994305e-57 -0.2744643 -1.4019208
4  3.209704e-42 -0.1049966 -1.3397542
```

When you are finished using a table, it can be removed from memory. Let's see how to do that in the next section.

Dropping a CAS Table

If a table is no longer needed, you can remove it by using the table.dropTable action.

```
> cas.table.dropTable(
    conn,
    name = 'data.iris',
    caslib = 'casuser'
    )
```

```
NOTE: Cloud Analytic Services dropped table data.iris from caslib
CASUSER(username).
```

In some of the previous listings, you might have seen data types that haven't previously been available in a SAS product. CAS has a much richer set of data types than traditional SAS does, so we'll look at those next.

CAS Data Types

When parsing the Iris CSV file in the previous examples, the character data was stored as a varchar (variable-width character) column. This means that the values are not padded with extra space like a fixed-width character column. This is just one of the new data types in CAS that aren't available in traditional SAS. Table 4.1 is a full list of the data types that are available in CAS tables.[1]

Table 4.1: Available Data Types in CAS Tables

Name	Description
Double	8-byte IEEE floating-point number
Int32	32-bit integer
Int64	64-bit integer
DecQuad	128-bit fixed-decimal value
DecSext	192-bit fixed-decimal value
Char	Fixed-width, blank-padded UTF-8 encoded character data
Varchar	Variable-width UTF-8 encoded character data
Binary	Fixed-width binary data
Varbinary	Variable-width binary data
Date	CAS date value (32-bit integer containing the number of days since January 1, 1960)
Time	CAS time value (64-bit integer containing the number of microseconds since midnight)
Datetime	CAS datetime value (64-bit integer containing the number of microseconds since January 1, 1960)

With all of these data types, it is possible to do much more targeted data processing.

Caslib and CAS Table Visibility

There are two scopes in CAS for caslibs and CAS tables: session (or local) and global. By default, both caslibs and CAS tables are created with session scope. That means that they are visible only in the session that created them. If you need to use them across sessions, you need to specify a parameter indicating the global scope. The table.addCaslib action uses the session parameter to indicate session or global visibility. The table.loadTable and table.addTable actions use the promote parameter to indicate whether the table should be promoted to the global scope. CAS tables can also be promoted after creation by using the table.promote action. However, note that the promotion of a table is possible only in a global caslib.

The Active Caslib

As mentioned previously, there is one caslib that is considered active at any point in time. To be active means that any table references in actions that are not accompanied by a caslib name are assumed to be in the active caslib. You can use the table.caslibInfo action to see which caslib is active. In the following output, the Casuser caslib is marked as the active caslib.

```
> cas.table.caslibInfo(conn)
```

```
$CASLibInfo

                 Name   Type                      Description
  0  CASUSER(username)   PATH   Personal File System Caslib
  1           Formats   PATH                    Format Caslib

                       Path Definition   Subdirs   Local   Active
  0           /u/username/                   1.0     0.0      1.0
  1  /bigdisk/lax/formats/                   1.0     0.0      0.0

       Personal   Hidden
  0         1.0      0.0
  1         0.0      0.0
```

Alternatively, you can use the sessionProp.getSessOpt action to query the caslib option.

```
> cas.sessionProp.getSessOpt(conn, name = 'caslib')
```

```
$caslib
[1] "CASUSER(username)"
```

The active caslib can be set at creation using the activeOnAdd parameter of table.addCaslib, or it can be set at any time using the sessionProp.setSessOpt action.

```
> cas.sessionProp.setSessOpt(conn, caslib = 'formats')
```

```
NOTE: 'Formats' is now the active caslib.
```

Regardless of which caslib is active, it's generally a good idea to always specify the caslib that you load a table from. This practice helps avoid errors where the active caslib gets changed at another point in the program.

Let's set the active caslib back to casuser before moving on.

```
> cas.sessionProp.setSessOpt(conn, caslib = 'casuser')
```

```
NOTE: 'CASUSER(username)' is now the active caslib.
```

With the information from this chapter so far, you should be able load basic data files into CAS. If you have some data that you want to analyze, you could jump ahead to the chapters that cover the topics that you are interested in. However, if you want to learn about other ways of getting data into CAS, go to the next section.

Uploading Data Files to CAS Tables

In the previous section, we showed how to load a data file from a file-based caslib. But what if you don't have direct access to the file system where the CAS server is running, or you just want to upload data from a file that exists on the client side? The easiest solution is to use the built-in data loading functions cas.read.csv, cas.read.jmp, cas.read.sas7bdat, cas.read.xlsx, cas.readRDS, and cas.read.table.

In its simplest form, the built-in data loading functions just take a CAS connection object and a filename as arguments. The type of data that is in the file is inferred from the file extension, and the CAS table name is automatically generated. At the time of this writing, the file types that are supported by the functions are CSV, JMP, sas7bdat, XLSX, and RDS. Note that when loading data using the functions, the table exists only in-memory. The file is not stored anywhere on the server. Let's load the Iris data set from a local file rather than from a file on the server. We use the cas.read.csv function as an example in this section. The other functions are similar.

```
> iris <- cas.read.csv(conn, '/u/username/data/iris.csv')
```

```
NOTE: Cloud Analytic Services made the uploaded file available as table IRIS in
caslib Formats.
```

```
> class(iris)
```

```
[1] "CASTable"
attr(,"package")
[1] "swat"
```

As you can see, the output from the cas.read.csv function is a CASTable object. We can now query for the table, the column, and data information as we did in the previous section.

```
> cas.table.columnInfo(iris)
```

```
$ColumnInfo

          Column  ID    Type  RawLength  FormattedLength  NFL  NFD
0   Sepal.Length   1  double          8               12    0    0
1    Sepal.Width   2  double          8               12    0    0
2   Petal.Length   3  double          8               12    0    0
3    Petal.Width   4  double          8               12    0    0
4        Species   5 varchar         10               10    0    0
```

Also, just as with the table.loadTable action, you can specify a casout parameter in order to assign a different table name or caslib.

```
> iris2 <- cas.read.csv(
    conn,
    '/u/username/data/iris.csv',
    casOut = list(name = 'iris2', caslib='casuser')
    )
```

```
NOTE: Cloud Analytic Services made the uploaded file available as table IRIS2 in
caslib CASUSER(username).
```

So far, we have been using the default data parsing parameters for our CSV file. The functions support various options for each of the supported file types. The parameters used to specify parsing options are similar to those of the popular read.csv function. We can use the Help from the cas.read.csv function to see what options are available. Here is a partial listing:

```
> ?cas.read.csv
```

Using the sep parameter that was previously described, you can parse a tab-delimited file rather than a comma-delimited file by changing the delimiter.

```
> iris_tsv <- cas.read.csv(
    conn,
    '/u/username/data/iris.tsv',
    sep = '\t',
    casOut = list(name = "iris_tsv", caslib='casuser')
    )
```

```
NOTE: Cloud Analytic Services made the uploaded file available as table IRIS_TSV in
caslib CASUSER(username).
```

```
> head(iris_tsv)
  Sepal.Length Sepal.Width Petal.Length Petal.Width Species
1          5.1         3.5          1.4         0.2       0
2          4.9         3.0          1.4         0.2       0
3          4.7         3.2          1.3         0.2       0
4          4.6         3.1          1.5         0.2       0
5          5.0         3.6          1.4         0.2       0
6          5.4         3.9          1.7         0.4       0
```

Now that we have seen how to upload local files, let's examine uploading data from URLs.

Uploading Data from URLs to CAS Tables

Uploading data from a URL works exactly the same way as uploading from a local file. The only difference is that you are specifying a URL as the second parameter of the cas.read.csv function.

```
> out <- cas.read.csv(conn, 'https://raw.githubusercontent.com/
sassoftware/sas-viya-programming/master/data/class.csv')
```

```
NOTE: Cloud Analytic Services made the uploaded file available as table CLASS in
caslib CASUSER(username).
```

```
> head(out)
     Name Sex Age Height Weight
1  Alfred   M  14   69.0  112.5
2   Alice   F  13   56.5   84.0
3 Barbara   F  13   65.3   98.0
4   Carol   F  14   62.8  102.5
```

```
5    Henry   M   14   63.5   102.5
6    James   M   12   57.3    83.0
```

Uploading Data from a data.frame to a CAS Table

If you already have your data in an R data frame and you want to upload it to a CAS table, the as.casTable function is the quickest and easiest way. In the case of data frames, you give the data frame object as the second argument to the as.casTable function rather than a filename. In the following example, we upload the built-in iris data frame to CAS.

```
# Quick check of data values
> class(iris)
[1] "data.frame"

> head(iris)
```

```
  Sepal.Length Sepal.Width Petal.Length Petal.Width Species
1          5.1         3.5          1.4         0.2  setosa
2          4.9         3.0          1.4         0.2  setosa
3          4.7         3.2          1.3         0.2  setosa
4          4.6         3.1          1.5         0.2  setosa
5          5.0         3.6          1.4         0.2  setosa
6          5.4         3.9          1.7         0.4  setosa
```

```
# Upload the DataFrame
> iris_ct <- as.casTable(conn, iris)
```

```
NOTE: Cloud Analytic Services made the uploaded file available as table IRIS in
caslib CASUSER(username).
```

```
> cas.table.fetch(conn,table = 'iris', to = 5)
```

```
$Fetch
  _Index_ Sepal.Length Sepal.Width Petal.Length Petal.Width   Species
1       1          5.1         3.5          1.4         0.2    setosa
2       2          4.9         3.0          1.4         0.2    setosa
3       3          7.7         3.0          6.1         2.3 virginica
4       4          4.7         3.2          1.3         0.2    setosa
5       5          4.6         3.1          1.5         0.2    setosa
```

You'll see that in the case of data.frame, the table name is always the same as the data.frame name by default. You can give it a different name using the casOut parameter. In the following example, we upload the iris data.frame again, and name the resulting table "iris_cas".

```
> iris_tbl <- as.casTable(conn, iris, casOut = 'iris_cas')
```

```
NOTE: Cloud Analytic Services made the uploaded file available as table IRIS_CAS in
caslib CASUSER(username).
```

Exporting CAS Tables to Other Formats

The data in CAS tables can be saved to many of the formats that are supported by the R data.frame's export methods. It accomplishes this by fetching the data and using the data.frame export methods to save the data. These functions include cas.saveRDS, cas.write.csv, cas.write.csv2, cas.write.table, and cas.write.xlsx. Keep in mind that all of these functions save the output on the client side. To save the data in a CAS table to a data source, you use the table.save CAS action.

The table.save action stores the data in the CAS table to the appropriate format for the caslib where the data is stored. The most basic form is the file-based caslib. You can store the data from a CAS table to either the CSV form or the SASHDAT form in file-based caslibs. To store the data, you simply specify the caslib to which the file should be saved, and you specify the path to the file in the name parameter.

```
> cas.table.save(iris_tbl, name='data/irisout.csv', caslib='casuser')
```

```
NOTE: Cloud Analytic Services created the file data/irisout.csv in caslib
CASUSER(username).
caslib
[1] "CASUSER(username)"

$name
[1] "data/irisout.csv"
```

```
> cas.table.save(iris_tbl, name='data/irisout.sashdat', caslib='casuser')
```

```
NOTE: Cloud Analytic Services created the file data/irisout.sashdat in caslib
CASUSER(username).
$caslib
[1] "CASUSER(username)"

$name
[1] "data/irisout.sashdat"
```

For caslibs with other data sources, see the SAS documentation for the options and parameters that are needed to export the data to a specific data source type.

Managing Caslibs

We have used file system path-based caslibs throughout this chapter. These are the most common forms of caslibs, but there are other types such as HDFS, DNFS, ESP, and Hadoop. Coverage of all of them is beyond the scope of this book, but we cover how to set up and manage a new caslib in this section.

Creating a Caslib

Suppose that you have a collection of data files in a directory called /research/data/ and you want to create a caslib that enables access to only the files (but not the subdirectories) in that directory. The caslib should be accessible by all sessions and should not be the active caslib.

Consulting the documentation for the table.addCaslib action, you'll see that we need to use the parameters subDirs, session, and activeOnAdd. We'll also add a description just to be complete. Here is the code:

```
> cas.table.addCaslib(
    conn,
    path='/research/data',
    caslib='research',
    description='Research Data',
    subDirs=FALSE,
    session=FALSE,
    activeOnAdd=FALSE
    )
```

```
NOTE: Cloud Analytic Services added the caslib 'research'.
$CASLibInfo

        Name   Type   Description    Path Definition
  0   research  PATH  Research Data    /research/data/

      Subdirs  Local  Active  Personal  Hidden
  0     0.0    0.0     0.0      0.0      0.0
```

As you can see from the output, we now have a new caslib that matches the parameters that we specified. Subdirs is set to 0, which indicates that subdirectories cannot be accessed. Local is set to 0, which means that it can be seen in any session. Note that this setting might require additional permissions for your user account. Active is set to 0, which means that it needs to be specified explicitly in action calls in conjunction with paths and table names.

Setting an Active Caslib

If you later decide that you want to make this caslib the active caslib, you can use the sessionProp.setSessOpt action with the caslib parameter.

```
> cas.sessionProp.setSessOpt(conn, caslib = 'research')
```

```
NOTE: 'research' is now the active caslib.
```

```
> cas.table.caslibInfo(conn, caslib='research')
```

```
$CASLibInfo

        Name   Type   Description    Path Definition    \
  1   research  PATH  Research Data    /research/data/

      Subdirs  Local  Active  Personal  Hidden
  1     0.0    0.0     1.0      0.0      0.0
```

Dropping a Caslib

If you want to remove a caslib, it's as simple as using the table.dropCaslib action.

```
> cas.table.dropCaslib(conn, caslib = 'research')
```

```
NOTE: 'CASUSER(username)' is now the active caslib.
NOTE: Cloud Analytic Services removed the caslib 'research'.
```

Notice that if you drop a caslib that was the active caslib, the active flag gets set on the previously active caslib. Also, if you drop a global caslib, multiple users might be affected.

Conclusion

We have covered a lot of ground in this chapter about getting your data into CAS. We started with learning what caslibs and CAS tables are. We then used the table.loadTable action to load data files that are stored on the server. From the client side, we used the built-in data loading functions cas.read.csv, cas.read.jmp, cas.read.sas7bdat, cas.read.xlsx, cas.readRDS, and cas.read.table on the CAS connection object to upload local files, URLs, and R data frames.

We also went through the process of creating and dropping a caslib so that you can create your own.

With all of these techniques, you now have the ability to get data into CAS in various ways and to organize it. In the next chapter, we describe the CASTable object.

---

[1] Support for data types might vary depending on your version of CAS.

Chapter 5: First Steps with the CASTable Object

A commonly used object in SWAT is the CASTable object. The CASTable object is the most important object in the SWAT package besides the CAS connection object. It encapsulates CAS table parameters and operations in one object. Also, it enables you to directly call CAS actions on the table object rather than always having to supply it as a parameter to the action. There are other more advanced features that we discuss in the next chapter.

In this chapter, we first look at managing CASTable parameters and running actions on the CASTable objects directly.

First Steps with the CASTable Object

Creating a CASTable Object

The first task that we need to do before we work with CASTable objects is to create a data table in CAS. Let's use one of the table.loadTable examples from the previous chapter that loads some data.

```
> out <- cas.table.loadTable(conn, path='data/iris.csv', caslib='casuser')
```

```
NOTE: Cloud Analytic Services made the file data/iris.csv available as table
DATA.IRIS in caslib CASUSER(username).
```

```
> out
$caslib
[1] "CASUSER(username)"

$tableName
[1] "DATA.IRIS"
```

We have mentioned previously that the result of CAS action is a list object. Therefore, any of the member names can be accessed using R's list syntax.

```
> out$caslib
```
```
[1] "CASUSER(username)"
```

```
> out[['tableName']]
```
```
[1] "DATA.IRIS"
```

In addition, the list class enables you to retrieve a slice of the list using the single bracket operator [ ].

```
> out['tableName']
```
```
$tableName
[1] "DATA.IRIS"
```

If you look at the output from the preceding code, you'll see that the CASTable object points to the DATA.IRIS table in the CASUSER(username) caslib. Now that you have a table loaded in CAS, you can wrap the table in a CASTable object. There is a built-in defCasTable function in the SWAT package that creates CASTable objects that are registered with the connection. So, to create a CASTable object that references our DATA.IRIS table, we do the following:

```
> # Create the CASTable object manually
> newtbl <- defCasTable(conn, 'data.iris')

> # Verify the result
> head(newtbl)
```

```
  Sepal.Length Sepal.Width Petal.Length Petal.Width     Species
1          5.1         3.5          1.4         0.2 Iris-setosa
2          4.9         3.0          1.4         0.2 Iris-setosa
3          4.7         3.2          1.3         0.2 Iris-setosa
4          4.6         3.1          1.5         0.2 Iris-setosa
5          5.0         3.6          1.4         0.2 Iris-setosa
6          5.4         3.9          1.7         0.4 Iris-setosa
```

The CASTable object is bound to the connection object that is specified as the first parameter in the defCasTable function. That means that any actions that are executed on the CASTable object also run in that session. It also means that any action sets that get loaded into that session can be used on the CASTable object.

We used the table.columnInfo and table.fetch actions frequently in the previous chapter. Each time, we specified both the connection and the table in the action call. Rather than doing that, you can execute the actions only using the CASTable object. In other words, instead of doing this:

```
> cas.table.columnInfo(conn, table = 'data.iris')
```

```
$ColumnInfo
        Column ID    Type RawLength FormattedLength NFL NFD
1 Sepal.Length  1  double         8              12   0   0
2  Sepal.Width  2  double         8              12   0   0
3 Petal.Length  3  double         8              12   0   0
4  Petal.Width  4  double         8              12   0   0
5      Species  5 varchar        15              15   0   0
```

```
> cas.table.fetch(conn, table = 'data.iris', to = 5)
$Fetch
  _Index_ Sepal.Length Sepal.Width Petal.Length Petal.Width
1       1          5.1         3.5          1.4         0.2
2       2          4.9         3.0          1.4         0.2
3       3          6.7         3.3          5.7         2.5
4       4          4.7         3.2          1.3         0.2
5       5          4.6         3.1          1.5         0.2
          Species
1     Iris-setosa
2     Iris-setosa
3  Iris-virginica
4     Iris-setosa
5     Iris-setosa
```

You can simply do this:

```
> cas.table.columnInfo(newtbl)
$ColumnInfo
        Column ID    Type RawLength FormattedLength NFL NFD
1 Sepal.Length  1  double         8              12   0   0
2  Sepal.Width  2  double         8              12   0   0
3 Petal.Length  3  double         8              12   0   0
4  Petal.Width  4  double         8              12   0   0
5      Species  5 varchar        15              15   0   0

> cas.table.fetch(newtbl, to = 5)
$Fetch
  _Index_ Sepal.Length Sepal.Width Petal.Length Petal.Width
1       1          5.1         3.5          1.4         0.2
2       2          4.9         3.0          1.4         0.2
3       3          6.7         3.3          5.7         2.5
4       4          4.7         3.2          1.3         0.2
5       5          4.6         3.1          1.5         0.2
          Species
1     Iris-setosa
2     Iris-setosa
3  Iris-virginica
4     Iris-setosa
5     Iris-setosa
```

As you can see, calling actions on a CASTable is much more concise and doesn't require you to know the names of the table or the caslib. We briefly showed you the summary action in the previous chapter as well. Executing the summary action on our table now appears as follows:

```
> cas.simple.summary(newtbl)
```

```
$Summary
        Column Min Max   N NMiss     Mean   Sum       Std      StdErr
1 Sepal.Length 4.3 7.9 150     0 5.843333 876.5 0.8280661 0.06761132
2  Sepal.Width 2.0 4.4 150     0 3.054000 458.1 0.4335943 0.03540283
3 Petal.Length 1.0 6.9 150     0 3.758667 563.8 1.7644204 0.14406432
4  Petal.Width 0.1 2.5 150     0 1.198667 179.8 0.7631607 0.06231181
```

```
         Var      USS        CSS       CV   TValue          ProbT   Skewness
1 0.6856935 5223.85 102.16833 14.17113 86.42537 3.331256e-129  0.3149110
2 0.1880040 1427.05  28.01260 14.19759 86.26430 4.374977e-129  0.3340527
3 3.1131794 2583.00 463.86373 46.94272 26.09020 1.994305e-57  -0.2744643
4 0.5824143  302.30  86.77973 63.66747 19.23659 3.209704e-42  -0.1049966
     Kurtosis
1 -0.5520640
2  0.2907811
3 -1.4019208
4 -1.3397542
```

Now that you see how easily this works, we can try a new action: correlation.

```
> cas.simple.correlation(newtbl)
```

```
$CorrSimple
      Variable   N     Mean    Sum    StdDev Minimum Maximum
1 Sepal.Length 150 5.843333 876.5 0.8280661     4.3     7.9
2  Sepal.Width 150 3.054000 458.1 0.4335943     2.0     4.4
3 Petal.Length 150 3.758667 563.8 1.7644204     1.0     6.9
4  Petal.Width 150 1.198667 179.8 0.7631607     0.1     2.5

$Correlation
      Variable Sepal.Length Sepal.Width Petal.Length Petal.Width
1 Sepal.Length    1.0000000  -0.1093692    0.8717542   0.8179536
2  Sepal.Width   -0.1093692   1.0000000   -0.4205161  -0.3565441
3 Petal.Length    0.8717542  -0.4205161    1.0000000   0.9627571
4  Petal.Width    0.8179536  -0.3565441    0.9627571   1.0000000
```

From these examples, you can see that any action that takes a table definition as an argument can be executed directly on the CASTable object.

In addition to the action interface, you can also call many of the R data.frame functions. For example, rather than using the table.columnInfo action to get column names, you can use the names function. We'll cover a few here. In the next chapter, we discuss the data.frame compatibility features in more detail.

```
> names(newtbl)
[1] "Sepal.Length" "Sepal.Width"  "Petal.Length" "Petal.Width"
[5] "Species"
```

The summary function also works the same way as in data.frame objects that are complete with the percentiles, include, and exclude options. Of course, with the power of CAS behind this, you can retrieve the statistics computed by the summary method on data sets that are much larger than those that are supported by a conventional R data.frame.

```
> summary(newtbl)
Sepal.Length     Sepal.Width     Petal.Length    Petal.Width     Species
Min.   :4.300    Min.   :2.000   Min.   :1.000   Min.   :0.100   Iris-setosa    :50
1st Qu.:5.100    1st Qu.:2.800   1st Qu.:1.600   1st Qu.:0.300   Iris-versicolor:50
Median :5.800    Median :3.000   Median :4.350   Median :1.300   Iris-virginica :50
Mean   :5.843    Mean   :3.054   Mean   :3.759   Mean   :1.199
3rd Qu.:6.400    3rd Qu.:3.300   3rd Qu.:5.100   3rd Qu.:1.800
Max.   :7.900    Max.   :4.400   Max.   :6.900   Max.   :2.500
```

All of these functions call CAS actions in the background and reformat the results to the familiar data.frame output types. So, if you are familiar with R data.frames, you should feel comfortable working with the CASTable objects this way.

Now that we have shown the basics of CASTable objects, let's dig deeper into the action and parameter interfaces.

Setting CASTable Parameters

So far, we have demonstrated how to set the table name and a caslib of CASTable. However, there are many more possible parameters to use on both input and output tables. To see the full listing of possible parameters, you can create a CASTable and check its attributes. Here is an example listing:

```
> attributes(newtbl)
```

```
$conn
CAS(hostname=server-name.mycompany.com, port=8777, username=username,
session=02c86e7b-250b-334a-8556-b1655d3edd19, protocol=http)

$tname
[1] "data.iris"

$caslib
[1] ""

$where
[1] ""

$orderby
list()

$groupby
list()

$gbmode
[1] ""

$computedOnDemand
[1] FALSE

$computedVars
[1] ""

$computedVarsProgram
[1] ""

$XcomputedVarsProgram
[1] ""

$XcomputedVars
[1] ""

$names
[1] "Sepal.Length" "Sepal.Width"  "Petal.Length" "Petal.Width"
[5] "Species"
```

```
$compcomp
[1] FALSE

$class
[1] "CASTable"
attr(,"package")
[1] "swat"
```

Let's create a CASTable object that includes a where parameter to subset the rows.

One way to think of a CASTable object is as a view of a table on the CAS server. Even though you subset rows, or perform other operations on a CASTable object, such as add orderby variables, and create computed columns, you do not modify the table on the server side at all. These parameters are simply stored on the CASTable object and are automatically sent as table parameters when actions are called on the CASTable object. The referenced CAS table is always the same, but any methods or CAS actions that are called on the CASTable object are performed using the view of the data from that object.

```
> iristbl <- defCasTable(
+    conn, 'data.iris',
+    caslib = 'casuser',
+    where = '"Sepal.Length"n > 6.8 and Species = "Iris-virginica"'
+    )
```

And this CASTable object can be used as an input table.

```
> cas.simple.summary(
+    iristbl,
+    casout = list(name = 'summout', caslib = 'casuser', promote = TRUE)
+    )
```

```
$OutputCasTables
            casLib    Name Rows Columns
1 CASUSER(username) summout    4      17
```

```
> outtbl = defCasTable(conn, 'summout')
> cas.table.fetch(outtbl)
$Fetch
  _Index_  _Column_ _Min_ _Max_ _NObs_ _NMiss_    _Mean_   _Sum_
1       1 Sepal.Length  6.90  7.90     15       0  7.360000 110.40
2       2  Sepal.Width  2.60  3.80     15       0  3.126667  46.90
3       3 Petal.Length  5.10  6.90     15       0  6.120000  91.80
4       4  Petal.Width  1.60  2.50     15       0  2.086667  31.30

       _Std_    _StdErr_        _Var_    _USS_      _CSS_        _CV_
1 0.3376389 0.08717798  0.11400000 814.14  1.5960000  4.587485
2 0.3534860 0.09126970  0.12495238 148.39  1.7493333 11.305524
3 0.5017114 0.12954132  0.25171429 565.34  3.5240000  8.197898
4 0.2416215 0.06238640  0.05838095  66.13  0.8173333 11.579305

         _T_          _PRT_  _Skewness_  _Kurtosis_
1 84.42499 2.332581e-20  0.01387515  -1.3790042
2 34.25744 6.662768e-15  0.89896300   0.2055943
3 47.24361 7.680656e-17 -0.35930456  -0.2101472
4 33.44746 9.278838e-15 -0.34630294  -0.2688732
```

In addition to setting parameters on the constructor, you can also set parameters on existing instances. These options can be set using: 1) the function interface, and 2) the attribute interface.

Managing Parameters Using the Attribute Interface

CASTable parameters can be changed using an attribute-style interface, as well as the previously described function-style interface.

```
> iristbl <- defCasTable(conn, 'data.iris', caslib = 'casuser')
> dim(iristbl)
[1] 150    5
> iristbl@where = '"Sepal.Length"n > 6.8 and Species = "Iris-virginica"'
> dim(iristbl)
[1] 15  5
```

If your parameters include lists, you can pass a list directly to a parameter:

```
> iristbl@groupby <- list("Species", "Sepal.Length")
```

Or, you can use list indexing syntax to set them individually:

```
> iristbl@groupby[[1]] <- "Species"
```

```
> iristbl@groupby[[2]] <- "Sepal.Length"
```

The two methods produce the same result.

```
> attributes(iristbl)
```

```
$conn
CAS(hostname=server-name.mycompany.com, port=8777, username=username,
session=02c86e7b-250b-334a-8556-b1655d3edd19, protocol=http)

$tname
[1] "data.iris"

$caslib
[1] "casuser"

$where
[1] "\"Sepal.Length\"n > 6.8 and Species = \"Iris-virginica\""

$orderby
list()

$groupby
$groupby[[1]]
[1] "Species"

$groupby[[2]]
[1] "Sepal.Length"


$gbmode
[1] ""
```

```
$computedOnDemand
[1] FALSE

$computedVars
[1] ""

$computedVarsProgram
[1] ""

$XcomputedVarsProgram
[1] ""

$XcomputedVars
[1] ""

$names
[1] "Sepal.Length" "Sepal.Width"  "Petal.Length" "Petal.Width"  "Species"

$compcomp
[1] FALSE

$class
[1] "CASTable"
attr(,"package")
[1] "swat"
```

Retrieving parameter values and deleting them also work with R's attribute syntax.

```
> iristbl@groupby
```

```
[[1]]
[1] "Species"

[[2]]
[1] "Sepal.Length"
```

```
> iristbl@groupby <- list()

> iristbl@where <- ''

> attributes(iristbl)
```

```
$conn
CAS(hostname=server-name.mycompany.com, port=8777, username=username,
session=02c86e7b-250b-334a-8556-b1655d3edd19, protocol=http)

$tname
[1] "data.iris"

$caslib
[1] "casuser"

$where
[1] ""

$orderby
list()
```

```
$groupby
list()

$gbmode
[1] ""

$computedOnDemand
[1] FALSE

$computedVars
[1] ""

$computedVarsProgram
[1] ""

$XcomputedVarsProgram
[1] ""

$XcomputedVars
[1] ""

$names
[1] "Sepal.Length" "Sepal.Width"  "Petal.Length" "Petal.Width"  "Species"

$compcomp
[1] FALSE

$class
[1] "CASTable"
attr(,"package")
[1] "swat"
```

The groupby parameter is set as a list in the previous example. You can also set it as a vector. Vectors are equivalent to lists as far as SWAT is concerned.

```
> iristbl@groupby <- c('Species', 'Sepal.Length')
```

And you can also set the computedVars and computedVarsProgram parameters to create computed columns. We haven't used computedVars or computedVarsProgram previously. The computedVarsProgram parameter is a string that contains SAS code to create the values for the computed columns. The computedVars parameter is a list of the variable names that are created by the computedVarsProgram parameter and that show up as computed columns in the table

```
> iristbl@computedVars <- c('Length.Factor')
> iristbl@computedVarsProgram <- 'Length.Factor = Sepal.Length * Petal.Length'
> attributes(iristbl)
$conn
CAS(hostname=server-name.mycompany.com, port=8777, username=username,
session=029b455a-6d3f-644a-b2f3-950a88fb2e07, protocol=http)

$tname
[1] "data.iris"

$caslib
[1] "casuser"
```

```
$where
[1] "\"Sepal.Length\"n > 6.8 and Species = \"Iris-virginica\""

$orderby
[1] ""

$groupby
[1] ""

$gbmode
[1] ""

$computedOnDemand
[1] FALSE

$computedVars
[1] "Length.Factor"

$computedVarsProgram
[1] "Length.Factor = Sepal.Length * Petal.Length"

$XcomputedVarsProgram
[1] ""

$XcomputedVars
[1] ""

$names
[1] "Sepal.Length" "Sepal.Width"  "Petal.Length" "Petal.Width"
[5] "Species"

$compcomp
[1] FALSE

$class
[1] "CASTable"
attr(,"package")
[1] "swat"


# Use the fetchvars= parameter to only fetch specified columns
> cas.table.fetch(iristbl, fetchvars = c('Sepal.Length', 'Petal.Length',
'Length.Factor'))
```

```
$Fetch
    _Index_ Sepal.Length Petal.Length Length.Factor
1        1          7.1          5.9         41.89
2        2          7.6          6.6         50.16
3        3          7.3          6.3         45.99
4        4          7.2          6.1         43.92
5        5          7.7          6.7         51.59
6        6          7.7          6.9         53.13
7        7          6.9          5.7         39.33
8        8          7.7          6.7         51.59
9        9          7.2          6.0         43.20
10      10          7.2          5.8         41.76
11      11          7.4          6.1         45.14
12      12          7.9          6.4         50.56
```

```
13      13      7.7      6.1      46.97
14      14      6.9      5.4      37.26
15      15      6.9      5.1      35.19
```

That covers just about anything you need to do with CASTable parameters. We'll show you how to materialize them in a real table in the server in the next section.

Materializing CASTable Parameters

We mentioned previously that CASTable objects are essentially client-side views of the data in a CAS table. Setting parameters on a CASTable object has no effect on the table in the server. Once you have created a CASTable with all of your computed columns and filters, you might want to materialize them on to the server as an in-memory table so that you can access it from other CASTable references. You can use the table.partition action to do this. Note that this is not the only use for the table.partition action, but it works in this case as well. Just as with table.loadTable, the table.partition action output has a casTable key that contains a reference to the new CASTable object.

```
> iristbl <- defCasTable(
+     conn, 'data.iris',
+     caslib = 'casuser',
+     where = '"Sepal.Length"n > 6.8 and Species = "Iris-virginica"'
+ )
> sub_iris <- cas.table.partition(iristbl)
> sub_iris
$averageShuffleWaitTime
[1] 0

$caslib
[1] "CASUSER(username)"

$maxShuffleWaitTime
[1] 0

$minShuffleWaitTime
[1] 1e+300

$rowsTransferred
[1] 0

$shuffleWaitTime
[1] 0

$tableName
[1] "_T_KY7CHW5V_CZEOCGIF_VTNJ3UYJ5T"

> sub_iris <- defCasTable(conn, sub_iris$tableName)
> cas.table.fetch(sub_iris)
$Fetch
  _Index_ Sepal.Length Sepal.Width Petal.Length Petal.Width      Species
1       1          7.1         3.0          5.9         2.1 Iris-virginica
2       2          7.6         3.0          6.6         2.1 Iris-virginica
3       3          7.3         2.9          6.3         1.8 Iris-virginica
4       4          7.2         3.6          6.1         2.5 Iris-virginica
5       5          7.7         3.8          6.7         2.2 Iris-virginica
6       6          7.7         2.6          6.9         2.3 Iris-virginica
7       7          6.9         3.2          5.7         2.3 Iris-virginica
```

```
8      8       7.7     2.8     6.7     2.0 Iris-virginica
9      9       7.2     3.2     6.0     1.8 Iris-virginica
10     10      7.2     3.0     5.8     1.6 Iris-virginica
11     11      7.4     2.8     6.1     1.9 Iris-virginica
12     12      7.9     3.8     6.4     2.0 Iris-virginica
13     13      7.7     3.0     6.1     2.3 Iris-virginica
14     14      6.9     3.1     5.4     2.1 Iris-virginica
15     15      6.9     3.1     5.1     2.3 Iris-virginica

> cas.simple.numRows(sub_iris)
```

```
$numrows
```

```
[1] 15
```

Conclusion

In this chapter, we introduced the CASTable object, and showed you various ways of setting parameters on CASTable instances. Depending on your coding style or how your parameters are being generated, you can choose the appropriate method for setting your action parameters. Now that we have seen the basics of the CASTable object, let's move on to advanced usage of the CASTable objects.

Chapter 6: Working with CAS Tables

In the previous chapter, we introduced CASTable objects. We covered the methods of setting parameters and calling actions on a CASTable object, but our coverage only included the basics. In this chapter we continue to work with CASTable objects on a variety of topics, including the data.frame APIs, exporting data, creating visualizations, fetching and indexing, creating temporary columns, and by-group processing.

Using CASTable Objects like a Data Frame

The R data.frame is a table of named columns each containing a vector of one type of data. They create the common data structure for statistical analyses in R packages, and also define interesting ways to index and operate on data sets. However, data.frames are limited to only holding data that can fit into the memory of your computer.

The CASTable object adopts many of these APIs and applies them to tables in CAS. This means that you can use the familiar data.frame API on CAS tables. The result types that are returned by data.frame functions on a CASTable object are, in most cases, the same as those that are returned by those functions on a data.frame. The primary difference is that the amount of data that is referenced by a CASTable object can be much greater than what can be handled on a single desktop machine.

In the following sections, we start with some functions and attributes that describe the CAS table itself, and then move on to functions for simple statistics and other areas of the data.frame API.

CAS Table Introspection

Basic information about CAS tables includes the names of the columns, the data types, and the number of rows. Until now, we've used the table.tableInfo and table.columnInfo CAS actions to get that information. However, we can also use the data.frame attributes and functions for the same purpose. Probably the most common function that is used to get this type of information is the names function. It simply lists the names of all of the columns in the table. We use the built-in Iris data in the data sets package for comparison.

```
> # check the class of iris data to make sure it is a data.frame instance
> class(iris)
```

```
[1] "data.frame"
```

```
> # Get the column names
> names(iris)
```

```
[1] "Sepal.Length" "Sepal.Width"  "Petal.Length" "Petal.Width"
[5] "Species"
```

```
> # Load iris data.frame into a CAS table
> tbl <- as.casTable(conn, iris)
```

```
NOTE: Cloud Analytic Services made the uploaded file available as table IRIS in
caslib CASUSER(username).
```

```
> # Show the class of tbl object
> class(tbl)
```

```
[1] "CASTable"
attr(,"package")
[1] "swat"
```

```
> names(tbl)
```

```
[1] "Sepal.Length" "Sepal.Width"  "Petal.Length" "Petal.Width"
[5] "Species"
```

In both cases, you see that the names function returns a vector object that contains the names of the columns.

To obtain the size or the shape of a CAS table, you can use the nrow and the dim functions just as in a data frame.

```
> nrow(tbl)
```

```
[1] 150
```

```
> dim(tbl)
```

```
[1] 150   5
```

For a quick summary of information about the entire table, you can use the attributes function.

```
> attributes(tbl)
```

```
$conn
CAS(hostname=server-name.mycompany.com, port=8777, username=username,
session=401486f5-6d36-b148-91f9-12bbd8f29e3f, protocol=http)

$tname
[1] "iris"

$caslib
[1] ""

$where
[1] ""

$orderby
[1] ""

$groupby
[1] ""

$gbmode
[1] ""

$computedOnDemand
[1] FALSE

$computedVars
[1] ""

$computedVarsProgram
[1] ""

$XcomputedVarsProgram
[1] ""

$XcomputedVars
[1] ""

$names
[1] "Sepal.Length" "Sepal.Width"  "Petal.Length" "Petal.Width"
[5] "Species"

$class
[1] "CASTable"
attr(,"package")
[1] "swat"
```

Finally, to get a sample of the data, you can use the head or tail functions. Keep in mind that CAS commonly uses distributed data storage that has no particular sort order, so the values that you get from the head and tail functions are not deterministic unless the table has a specified sort order, which is covered later in this chapter.

```
> head(tbl, n = 3L)
```

```
  Sepal.Length Sepal.Width Petal.Length Petal.Width   Species
1          5.1         3.5          1.4         0.2    setosa
2          4.9         3.0          1.4         0.2    setosa
3          7.7         3.0          6.1         2.3 virginica
```

```
> tail(tbl, n = 3L)
```

```
  Sepal.Length Sepal.Width Petal.Length Petal.Width   Species
1          6.4         2.8          5.6         2.2 virginica
2          6.3         2.8          5.1         1.5 virginica
3          6.1         2.6          5.6         1.4 virginica
```

As you can see from the preceding code, the head and tail functions accept the n parameter to indicate the number of records to retrieve (the default is 6).

Although this part of the data.frame API is fairly simple, it already shows how easy working with CASTable objects can be. Let's move on to computing some basic statistics.

Computing Simple Statistics

A common data.frame function that is used to learn about the characteristics of your data is the summary function. Here is a simple example of getting basic summary statistics from an R data.frame and a CASTable, both using the summary function.

```
# Run the summary function on a data.frame
> summ <- summary(iris)
> summ
```

```
  Sepal.Length    Sepal.Width     Petal.Length    Petal.Width
 Min.   :4.300   Min.   :2.000   Min.   :1.000   Min.   :0.100
 1st Qu.:5.100   1st Qu.:2.800   1st Qu.:1.600   1st Qu.:0.300
 Median :5.800   Median :3.000   Median :4.350   Median :1.300
 Mean   :5.843   Mean   :3.057   Mean   :3.758   Mean   :1.199
 3rd Qu.:6.400   3rd Qu.:3.300   3rd Qu.:5.100   3rd Qu.:1.800
 Max.   :7.900   Max.   :4.400   Max.   :6.900   Max.   :2.500
       Species
 setosa    :50
 versicolor:50
 virginica :50
```

```
> class(summ)
```

```
[1] "table"
```

```
# Run the summary function on a CASTable
> cassumm <- summary(tbl)
> cassumm
```

```
   Sepal.Length      Sepal.Width      Petal.Length      Petal.Width
 Min.   :4.300    Min.   :2.000    Min.   :1.000    Min.   :0.100
 1st Qu.:5.100    1st Qu.:2.800    1st Qu.:1.600    1st Qu.:0.300
 Median :5.800    Median :3.000    Median :4.350    Median :1.300
 Mean   :5.843    Mean   :3.057    Mean   :3.758    Mean   :1.199
 3rd Qu.:6.400    3rd Qu.:3.300    3rd Qu.:5.100    3rd Qu.:1.800
 Max.   :7.900    Max.   :4.400    Max.   :6.900    Max.   :2.500
         Species
 setosa    :50
 versicolor:50
 virginica :50
```

```
> class(cassumm)
```

```
[1] "table"
```

As you can see, in both cases, we run the summary function and get back the same results even in the same data type. This might not seem impressive for only 150 records, but how about running it on a larger collection? Our table contains 70 million rows and 46 columns, 35 of which are numeric. Running the summary function on our grid of 48 nodes requires only seconds.

```
> cas.table.tableInfo(conn, 'MEGACORP5M')
```

```
$TableInfo
          Name                              Label      Rows Columns
IndexedColumns
1 MEGACORP5M MegaCorp Production Data - 70 Million rows 70732833      46
0
  Encoding       CreateTimeFormatted          ModTimeFormatted
AccessTimeFormatted
1    latin1 2018-06-01T15:58:01-04:00 2018-06-01T15:58:01-04:00 2018-06-01T16:03:14-
04:00
  JavaCharSet CreateTime    ModTime AccessTime Global Repeated View
SourceName
1   ISO8859_1 1843502281 1843502281 1843502594      0        0     0
megacorp5m.sashdat
  SourceCaslib Compressed Creator Modifier    SourceModTimeFormatted SourceModTime
1      yourcaslib     0 username    2017-02-02T12:42:42-04:00    1801672962
```

```
> mega = defCasTable(conn,'MEGACORP5M')
> system.time({ result = summary(mega) })
Selecting by Frequency
   user   system elapsed
   0.22     0.00    8.48
```

```
> result
```

```
      Date           DateByYear        DateByMonth
 Min.   : 7305   Min.   : 7305    Min.   : 7305
 1st Qu.:12460   1st Qu.:12419    1st Qu.:12450
 Median :14605   Median :14245    Median :14579
```

```
Mean    :14389    Mean    :14209    Mean    :14375
3rd Qu.:16517     3rd Qu.:16437     3rd Qu.:16496
Max.    :18992    Max.    :18628    Max.    :18962
...

StateLongitude        StateLatitude       CityLongitude
Min.    :-119.99      Min.    :31.05      Min.    : -122.42
1st Qu.: -99.68       1st Qu.:31.17       1st Qu.:  -97.40
Median : -92.55       Median :32.62       Median :  -92.29
Mean    : -96.62      Mean    :35.09      Mean    :  -96.00
3rd Qu.: -86.74       3rd Qu.:37.27       3rd Qu.:  -88.04
Max.    : -71.55      Max.    :47.27      Max.    :  -71.46
                                          NA's    :612883.00
 CityLatitude
Min.    :   27.80
1st Qu.:   30.69
Median :   33.51
Mean    :   34.37
3rd Qu.:   37.77
Max.    :   47.61
NA's    :612883.00
```

Now that we've seen that we can analyze both small and large quantities of data using familiar data.frame functions, let's see what else it can do.

Besides the summary function, the SWAT package also provides access to the individual statistics through the functions cas.count, cas.mean, cas.std, cas.max, cas.nmiss, cas.sum, cas.stderr, cas.var, cas.uss, cas.cv, cas.tvalue, and cas.probt. These functions return a casDataFrame object that contains the values for each variable.

```
> cas.count(tbl)
```

```
          Column    N
1 Sepal.Length 150
2  Sepal.Width 150
3 Petal.Length 150
4  Petal.Width 150
```

```
> cas.mean(tbl)
```

```
          Column      Mean
1 Sepal.Length 5.843333
2  Sepal.Width 3.057333
3 Petal.Length 3.758000
4  Petal.Width 1.199333
```

```
> cas.var(tbl)
```

```
          Column       Var
1 Sepal.Length 0.6856935
2  Sepal.Width 0.1899794
3 Petal.Length 3.1162779
4  Petal.Width 0.5810063
```

At the time of this writing, not all of the R data.frame functions were supported by the CASTable object. As development of the SWAT package continues, more of the data.frame functions are likely to become available.

Now that we have seen some basic data.frame operations working on CASTable objects, let's move on to some plotting operations.

Creating Plots from CASTable Data

The easiest way to generate a plot of data in a CASTable is to use the plot function. The CASTable objects support all of the options in the plot function as the data.frame object because they use the same code. Plotting is done on the client side, so all of the data that is being visualized must be downloaded first. The plot function is implemented for CASTable objects, which downloads the data automatically. To prevent downloading very large data sets to the client, only 10,000 rows are randomly sampled and downloaded if the data set has more than 10,000 rows. The plot function is convenient, but for iterative work that might require tuning your plots, you probably want to avoid downloading data from the server at each iteration. You can do it by bringing the data back to a local data.frame and using the plot function on it directly.

We'll use our original Iris data set for this example. The following code creates the scatter plot shown in Figure 6.1:

```
> plot(tbl$Sepal.Length, tbl$Sepal.Width, pch = 16, type = 'p',
+       col = 'blue', main = "Plot on a CASTable")
```

Figure 6.1: Plot on a CAS Table

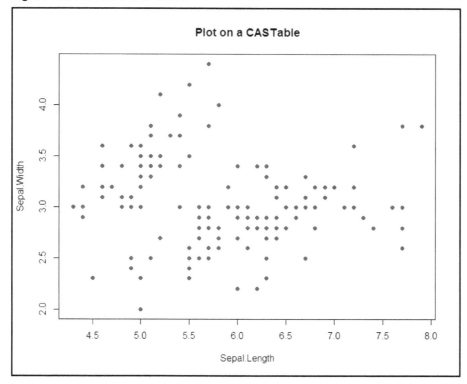

Of course, you can display subsets of observations as well using the standard R data.frame syntax. The following code generates Figure 6.2, which shows scatter plots of the three Iris species, displayed side by side.

```
> par(mfrow=c(1,3))
>
> tbl_tmp <- tbl[tbl$Species == 'setosa',]
> plot(tbl_tmp$Sepal.Length, tbl_tmp$Sepal.Width, pch = 16, type = 'p',
+      col = 'red', main = "setosa")
>
> tbl_tmp <- tbl[tbl$Species == 'versicolor',]
> plot(tbl_tmp$Sepal.Length, tbl_tmp$Sepal.Width, pch = 17, type = 'p',
+      col = 'green', main = "versicolor")
>
> tbl_tmp <- tbl[tbl$Species == 'virginica',]
> plot(tbl_tmp$Sepal.Length, tbl_tmp$Sepal.Width, pch = 18, type = 'p',
+      col = 'blue', main = "virginica")
```

Figure 6.2: Scatter Plots of the Three Iris Species

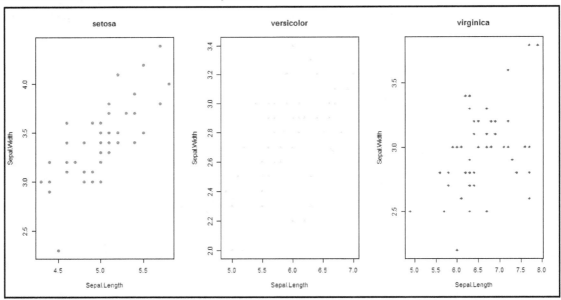

Sorting, Data Selection, and Iteration

Some of the most basic data operations deal with sorting rows, selecting data based on labels or expressions, and iterating over rows or columns of a table. The following sections introduce techniques for accomplishing each data operation using the CASTable object.

Fetching Data with a Sort Order

As mentioned previously, data in a CAS table might be distributed across a grid, or it can be accessible in ways that do not specify a particular row order. This means that anytime you fetch data from CAS, you can't depend on the order unless you specify a sort order when retrieving it. A sort order can be specified with the sortby parameter on the table.fetch action. Or you can specify an orderby parameter inside a CASTable object. The head function fetches and prints out the smallest rows of data, and the tail function fetches and prints out the largest rows. The order is ascending only. You need to use the table.fetch action if you want descending order.

There are two ways to specify an orderby parameter. You can either specify it when you define a CASTable object or use the orderby attribute in a CASTable object after creation of a CASTable object. Assuming that the "iris" table is available on the CAS server, you can use the following code to apply a sort order to a CASTable definition:

```
> sorttbl <- defCasTable(conn, 'iris', orderby = 'Sepal.Length')
```

Or you can use this code:

```
> sorttbl <- defCasTable(conn, 'iris')
> sorttbl@orderby <- 'Sepal.Length'
```

We now have a CASTable object with the sort order applied as indicated by the orderby parameter. To clarify, this sorting does nothing on the server side at this point. In fact, nothing was sent to the server when this operation was invoked. The orderby parameter simply binds a sort order to the CASTable object that gets used anytime that the CASTable object is used to retrieve data from that CAS table.

We can now use the head or tail functions to retrieve a sample of the data in the specified sort order.

```
> head(sorttbl, n = 10L)
```

	Sepal.Length	Sepal.Width	Petal.Length	Petal.Width	Species
1	4.3	3.0	1.1	0.1	setosa
2	4.4	3.2	1.3	0.2	setosa
3	4.4	3.0	1.3	0.2	setosa
4	4.4	2.9	1.4	0.2	setosa
5	4.5	2.3	1.3	0.3	setosa
6	4.6	3.6	1.0	0.2	setosa
7	4.6	3.1	1.5	0.2	setosa
8	4.6	3.2	1.4	0.2	setosa
9	4.6	3.4	1.4	0.3	setosa
10	4.7	3.2	1.6	0.2	setosa

```
> tail(sorttbl, n = 10L)
```

	Sepal.Length	Sepal.Width	Petal.Length	Petal.Width	Species
1	7.2	3.0	5.8	1.6	virginica
2	7.2	3.2	6.0	1.8	virginica
3	7.3	2.9	6.3	1.8	virginica
4	7.4	2.8	6.1	1.9	virginica
5	7.6	3.0	6.6	2.1	virginica
6	7.7	3.0	6.1	2.3	virginica
7	7.7	2.8	6.7	2.0	virginica
8	7.7	2.6	6.9	2.3	virginica
9	7.7	3.8	6.7	2.2	virginica
10	7.9	3.8	6.4	2.0	virginica

Now that we have seen how to apply a sort order to a CASTable, we can go to the next set of data.frame features that pertain to iteration.

Iterating through Columns

Iterating through the column names is the most basic form of iteration on a CASTable object. The following example iterates through column names and prints them out.

```
> for (col in names(sorttbl)){
+     print(col)
+ }
```

```
[1] "Sepal.Length"
[1] "Sepal.Width"
[1] "Petal.Length"
[1] "Petal.Width"
[1] "Species"
```

Iterating through rows is currently not supported.

Now that we've seen how to iterate through our data, let's look at how we can retrieve items from the data.

Techniques for Indexing and Selecting Data

Indexing and selecting data in an R data.frame comes in many forms. You can use the $ operator and the double brackets operator [[ ]] to select a single column. You can also use the single bracket operator [ ] to select multiple columns. The single bracket operator with a comma inside [,] enables you to select both rows and columns.

Selecting Columns by Label and Position

Probably the most common way that you see the data.frame being indexed is directly on the data.frame itself. Data.frame objects (and similarly CASTables objects) enable you to select columns either by name or by index. To select a single column, you simply index the CASTable using the name of the column.

```
> sorttbl@orderby <- 'Sepal.Width'
> col <- sorttbl['Sepal.Width']
> head(col)
```

```
   Sepal.Width
1       2.0
2       2.2
3       2.2
4       2.2
5       2.3
6       2.3
```

The object that we get back is still a CASTable object with a single column. Technically, CAS doesn't support stand-alone columns. If you use double brackets sorttbl[['Sepal.Width]] or sorttbl[,"Sepal.Width"], the returned object is still a CASTable object that references a single column of the table. As you can see, when you retrieve a column from a CASTable, the resulting CASTable with a single column takes with it all table and sort information that was on the parent CASTable.

An alternative way of selecting a column is using the R "$" operator. You can specify the name of the column on the CASTable in order to have it return a column.

```
> col <- sorttbl$Sepal.Width
> head(col)
```

```
  Sepal.Width
1         2.0
2         2.2
3         2.2
4         2.2
5         2.3
6         2.3
```

Although this way of accessing columns is prettier than the bracket syntax, note that the result is still a CASTable object with a single column, while the R "$" operator returns a vector. Although this syntax is appropriate for interactive use, it's much safer to write programs using the bracket syntax, because the bracketed syntax allows for column names that contain special characters that are not allowed by the $ operator.

In addition to retrieving a single column from a CASTable, we can also retrieve subsets of a table. To do this, you use a vector of column names as the index key.

```
> widths <- sorttbl[c('Sepal.Width', 'Petal.Width', 'Species')]
> class(widths)
```

```
[1] "CASTable"
attr(,"package")
[1] "swat"
```

```
> head(widths)
```

```
  Sepal.Width Petal.Width    Species
1         2.0         1.0 versicolor
2         2.2         1.0 versicolor
3         2.2         1.5  virginica
4         2.2         1.5 versicolor
5         2.3         1.3 versicolor
6         2.3         1.3 versicolor
```

If you specify a list of column names (or even a list containing a single column name), you'll get back a new CASTable object that includes only the referenced columns. The data in this table is still the same as the data in the parent table. The local CASTable object simply references the columns that it needs.

Of course, now that we have this new table, we can use any of the previously described CASTable methods on it (including the CAS action and the data.frame functions).

```
> summary(widths)
```

```
  Sepal.Width      Petal.Width          Species
 Min.   :2.000   Min.   :0.100   setosa    :50
 1st Qu.:2.800   1st Qu.:0.300   versicolor:50
 Median :3.000   Median :1.300   virginica :50
 Mean   :3.057   Mean   :1.199
 3rd Qu.:3.300   3rd Qu.:1.800
 Max.   :4.400   Max.   :2.500
```

```
> cas.table.columnInfo(widths)
```

```
$ColumnInfo
        Column ID    Type RawLength FormattedLength NFL NFD
1 Sepal.Width  2  double         8              12   0   0
2 Petal.Width  4  double         8              12   0   0
3     Species  5 varchar        10              10   0   0
```

In addition to basic column selection such as the examples shown here, you can also use the [,] of data.frame objects to select columns in other ways. That's covered in the next section.

Selecting Data by Label and Position

Although data.frames enable you to select rows using a comma inside the single bracket operator [,], CAS tables do not have the concept of a row label or an index. The operator still exists on CASTable objects, but they are limited to column selection only.

Let's use our sorted table from the previous example to select some data points.

```
> sorttbl@orderby <- 'Sepal.Width'
```

The [,] operator on CASTable objects can select only columns either by name or by index. Because of this, you should leave the row index portion empty. Again, note that CAS doesn't support stand-alone columns. Therefore, the returned object is still a CASTable object that references a single column of the table, not a vector object.

```
> # Select the petal_width column
> head(sorttbl[,"Petal.Width"])
```

```
  Petal.Width
1         1.0
2         1.0
3         1.5
4         1.5
5         1.3
6         1.3
```

```
> # Select a list of columns
> head(sorttbl[,c("Petal.Width","Sepal.Length")])
```

	Petal.Width	Sepal.Length
1	1.0	5.0
2	1.0	6.0
3	1.5	6.0
4	1.5	6.2
5	1.3	5.5
6	1.3	6.3

You can also select a single column or multiple columns using an integer index.

```
> head(sorttbl[,3])
```

	Petal.Length
1	3.5
2	4.0
3	5.0
4	4.5
5	4.0
6	4.4

```
> head(sorttbl[,1:3])
```

	Sepal.Length	Sepal.Width	Petal.Length
1	5.0	2.0	3.5
2	6.0	2.2	4.0
3	6.0	2.2	5.0
4	6.2	2.2	4.5
5	5.5	2.3	4.0
6	6.3	2.3	4.4

```
> head(sorttbl[,c(4,2)])
```

	Petal.Width	Sepal.Width
1	1.0	2.0
2	1.0	2.2
3	1.5	2.2
4	1.5	2.2
5	1.3	2.3
6	1.3	2.3

Although the preceding indexing methods can be useful, they are somewhat limited because the nature of distributed data prevents selection by row index. However, there is a more powerful indexing feature on CASTable objects that allows for various forms of advanced selection of data.

Dynamic Data Selection

Rather than using fixed values of rows and columns to select data, we can create conditions that are based on the data in the table to determine which rows to select. The specification of conditions is done using the same syntax as that used by data.frame objects. CASTable objects support R's various comparison operators and build a filter that subsets the rows in the table. You can then use the result of that comparison to index into a CASTable. It sounds much more complicated than it is, so let's look at an example.

If we want to get a CASTable that contains only values where Petal.Length is greater than 6.5, we can use the following expression to create our filter:

```
> expr <- sorttbl$Petal.Length > 6.5
```

This expression creates a computed column that is used in a where expression on the CASTable. This expression can then be used as an index value for a CASTable. Indexing this way essentially creates a Boolean mask. Wherever the expression values are true, the rows of the table are returned. Wherever the expression is false, the rows are filtered out.

```
> newtbl <- sorttbl[expr,]
> head(newtbl)
```

	Sepal.Length	Sepal.Width	Petal.Length	Petal.Width	Species
1	7.7	2.6	6.9	2.3	virginica
2	7.7	2.8	6.7	2.0	virginica
3	7.6	3.0	6.6	2.1	virginica
4	7.7	3.8	6.7	2.2	virginica

These two steps are commonly entered on one line.

```
> newtbl <- sorttbl[sorttbl$Petal.Length > 6.5,]
> head(newtbl)
```

	Sepal.Length	Sepal.Width	Petal.Length	Petal.Width	Species
1	7.7	2.6	6.9	2.3	virginica
2	7.7	2.8	6.7	2.0	virginica
3	7.6	3.0	6.6	2.1	virginica
4	7.7	3.8	6.7	2.2	virginica

We can further filter rows out by indexing another comparison expression.

```
> newtbl2 <- newtbl[newtbl$Petal.Width < 2.2,]
> head(newtbl2)
```

	Sepal.Length	Sepal.Width	Petal.Length	Petal.Width	Species
1	7.7	2.8	6.7	2.0	virginica
2	7.6	3.0	6.6	2.1	virginica

Comparisons can be joined using the bitwise comparison operators & (and) and | (or). You must be careful with these operators though due to operator precedence. Bitwise comparison has a lower precedence than comparisons such as greater-than and less-than, but it is still safer to enclose your comparisons in parentheses.

```
> newtbl3 <- sorttbl[(sorttbl$Petal.Length > 6.5) & (sorttbl$Petal.Width < 2.2),]
> head(newtbl3)
```

	Sepal.Length	Sepal.Width	Petal.Length	Petal.Width	Species
1	7.7	2.8	6.7	2.0	virginica
2	7.6	3.0	6.6	2.1	virginica

In all cases, we are not changing anything about the underlying data in CAS. We are simply constructing a query that is executed with the CASTable when it is used as the parameter in a CAS action. You can see what is happening behind the scenes by displaying the attributes of the resulting CASTable objects.

```
> attributes(newtbl3)
```

```
$conn
CAS(hostname=server-name.mycompany.com, port=8777, username=username,
session=11ed56e2-f9dd-9346-8d01-44a496e68880, protocol=http)

$tname
[1] "iris"

$caslib
[1] ""

$where
[1] "((\"Petal.Length\"n > 6.5) AND (\"Petal.Width\"n < 2.2))"

$orderby
[1] "Sepal.Width"

$groupby
[1] ""

$gbmode
[1] ""

$computedOnDemand
[1] FALSE

$computedVars
[1] ""

$computedVarsProgram
[1] ""

$XcomputedVarsProgram
[1] ""

$XcomputedVars
[1] ""

$names
[1] "Sepal.Length" "Sepal.Width"  "Petal.Length" "Petal.Width"
[5] "Species"

$class
[1] "CASTable"
attr(,"package")
[1] "swat"
```

You can also do mathematical operations on columns with constants or on other columns within your comparisons.

```
> iris[(iris$Petal.Length + iris$Petal.Width) * 2 > 17.5,]
    Sepal.Length Sepal.Width Petal.Length Petal.Width    Species
118          7.7         3.8          6.7         2.2 virginica
119          7.7         2.6          6.9         2.3 virginica
```

The list of supported operators is shown in the following table:

Operator	Numeric Data	Character Data
+ (add)	✓	
- (subtract)	✓	
* (multiply)	✓	
/ (divide)	✓	
%% (modulo)	✓	
%/% (integer division)	✓	
^ (power)	✓	

The supported comparison operators are shown in the following table.

Operator	Numeric Data	Character Data
== (equality)	✓	✓
!= (inequality)	✓	✓
< (less than)	✓	✓
> (greater than)	✓	✓
<= (less than or equal to)	✓	✓
>= (greater than or equal to)	✓	✓

As you can see in the preceding tables, you can do comparisons on character columns as well. In the following example, all of the rows in which Species is equal to "virginica" are selected, and saved to a new CASTable object virginica. Note that in this case, data is still not duplicated.

```
> tbl <- defCasTable(conn, 'iris')
> virginica <- tbl[tbl$Species == 'virginica',]
> dim(virginica)
```

```
[1] 50  5
```

```
> head(virginica)
```

```
  Sepal.Length Sepal.Width Petal.Length Petal.Width    Species
1          7.7         3.0          6.1         2.3 virginica
2          6.3         3.4          5.6         2.4 virginica
3          6.4         3.1          5.5         1.8 virginica
4          6.0         3.0          4.8         1.8 virginica
5          6.9         3.1          5.4         2.1 virginica
6          6.7         3.1          5.6         2.4 virginica
```

It's easy to create powerful filters that are executed in CAS while still using the R syntax. However, the similarities to data.frame don't end there. CASTable objects can also create computed columns and by groups using similar techniques. Those topics are covered in the next section.

Data Wrangling on the Fly

There's a certain amount of "data wrangling" that is required to prepare your data for reporting or modeling. Data wrangling refers to data manipulation, including the transformation of data from one format to another format, by-group processing, and data aggregation. Such manipulations on large data sets often become bottlenecks to understanding your data, because most of the steps to manipulate data require that you create a new copy of the data or that you move data around the nodes of your cluster. In CAS, several ways are provided to speed up the data wrangling steps, such as temporary computed columns and group-by processing. Neither requires copying data or moving data.

Creating Computed Columns

In addition to using the column-based expressions to filter a CASTable object, you can also use them to build computed columns. Instead of comparing the result of an expression to another value, you simply set it to a column on the CASTable object. Here is an example that adds the Sepal.Length and the Sepal.Width and then multiplies the result by 2. The result is a new column in the CASTable that we set to Sepal.Factor.

```
> sorttbl['Sepal.Factor'] <- (sorttbl$Sepal.Length+sorttbl$Sepal.Width)*2
> head(sorttbl)
```

```
  Sepal.Length Sepal.Width Petal.Length Petal.Width    Species
1          5.0         2.0          3.5         1.0 versicolor
2          6.0         2.2          4.0         1.0 versicolor
3          6.0         2.2          5.0         1.5  virginica
4          6.2         2.2          4.5         1.5 versicolor
5          5.5         2.3          4.0         1.3 versicolor
6          6.3         2.3          4.4         1.3 versicolor
  Sepal.Factor
1         14.0
2         16.4
3         16.4
4         16.8
5         15.6
6         17.2
```

The Sepal.Factor column is a computed column that is not present in the original CAS table. It is a temporary column generated "on the fly" when an action is executed. It exists only for the duration of the action, and is not visible to any other action. Computed columns are efficient because you can use computed columns to dynamically filter or expand the data without actually duplicating the data to create those extra filters or columns. You can create computed columns from other computed columns as well. In the following code, we'll use our previously computed column as part of the expression for a new computed column.

```
> sorttbl['Total.Factor'] <- sorttbl$Sepal.Factor + sorttbl$Petal.Width +
+    sorttbl$Petal.Length
> head(sorttbl)
```

```
  Sepal.Length Sepal.Width Petal.Length Petal.Width    Species
1          5.0         2.0          3.5         1.0 versicolor
2          6.0         2.2          4.0         1.0 versicolor
3          6.0         2.2          5.0         1.5  virginica
4          6.2         2.2          4.5         1.5 versicolor
5          5.5         2.3          4.0         1.3 versicolor
6          6.3         2.3          4.4         1.3 versicolor
```

	Sepal.Factor	**Total.Factor**
1	14.0	**18.5**
2	16.4	**21.4**
3	16.4	**22.9**
4	16.8	**22.8**
5	15.6	**20.9**
6	17.2	**22.9**

The output from the head function shows that the extra columns are now available on the table. You can even create computed columns that are based on constants or R expressions (evaluated on the client) rather than on data in the table.

```
> sorttbl['names'] <- 'sepal / petal'
> head(sorttbl)
```

	Sepal.Length	Sepal.Width	Petal.Length	Petal.Width	Species
1	5.0	2.0	3.5	1.0	versicolor
2	6.0	2.2	4.0	1.0	versicolor
3	6.0	2.2	5.0	1.5	virginica
4	6.2	2.2	4.5	1.5	versicolor
5	5.5	2.3	4.0	1.3	versicolor
6	6.3	2.3	4.4	1.3	versicolor

	Sepal.Factor	Total.Factor	names
1	14.0	18.5	**sepal / petal**
2	16.4	21.4	**sepal / petal**
3	16.4	22.9	**sepal / petal**
4	16.8	22.8	**sepal / petal**
5	15.6	20.9	**sepal / petal**
6	17.2	22.9	**sepal / petal**

You can use any of the numeric, string, and comparison methods that are described in the filtering section to construct computed columns as well. In the following example, a new column is created that has Boolean values indicating whether Species is equal to "virginica".

```
> tbl = defCasTable(conn, 'iris')
> tbl$is.virginica  <- (tbl$Species == 'virginica')
> head(tbl)
```

	Sepal.Length	Sepal.Width	Petal.Length	Petal.Width	Species	is.virginica
1	5.1	3.5	1.4	0.2	setosa	0
2	4.9	3.0	1.4	0.2	setosa	0
3	7.7	3.0	6.1	2.3	virginica	1
4	4.7	3.2	1.3	0.2	setosa	0
5	4.6	3.1	1.5	0.2	setosa	0
6	6.3	3.4	5.6	2.4	virginica	1

Now that we have seen that we can filter our data and create computed columns, let's look at the topic of grouping data by data values.

By-Group Processing

Many analytic actions and functions support grouping by values in one or more columns in a table. There are a couple of ways to apply by grouping in CASTable objects.

The most direct way of applying by grouping to a CASTable object is to use the groupby attribute. We'll start with the Iris table for this example.

```
> tbl <- as.casTable(conn, iris)
```

```
NOTE: Cloud Analytic Services made the uploaded file available as table IRIS in
caslib CASUSER(username).
```

```
> grptbl <- tbl
> grptbl@groupby <- 'Species'
> attributes(grptbl)
```

```
$conn
CAS(hostname=server-name.mycompany.com, port=8777, username=username,
session=60c6e0fc-d690-ea48-9dbc-9692e7205455, protocol=http)

$tname
[1] "iris"

$caslib
[1] ""

$where
[1] ""

$orderby
[1] ""

$groupby
[1] "Species"

$gbmode
[1] ""

$computedOnDemand
[1] FALSE

$computedVars
[1] ""

$computedVarsProgram
[1] ""

$XcomputedVarsProgram
[1] ""

$XcomputedVars
[1] ""

$names
[1] "Sepal.Length" "Sepal.Width"  "Petal.Length" "Petal.Width"
[5] "Species"

$class
[1] "CASTable"
attr(,"package")
[1] "swat"
```

Now that we have some grouping variables set, we'll run the summary action.

```
> cas.simple.summary(grptbl, subset = c("min","max"))
```

```
$ByGroup1.Summary
        Column Min Max
1 Sepal.Length 4.3 5.8
2  Sepal.Width 2.3 4.4
3 Petal.Length 1.0 1.9
4  Petal.Width 0.1 0.6

$ByGroup2.Summary
        Column Min Max
1 Sepal.Length 4.9 7.0
2  Sepal.Width 2.0 3.4
3 Petal.Length 3.0 5.1
4  Petal.Width 1.0 1.8

$ByGroup3.Summary
        Column Min Max
1 Sepal.Length 4.9 7.9
2  Sepal.Width 2.2 3.8
3 Petal.Length 4.5 6.9
4  Petal.Width 1.4 2.5

$ByGroupInfo
     Species    Species_f       _key_
1     setosa       setosa setosa
2 versicolor   versicolor versicolor
3  virginica    virginica virginica
```

You see that we now get multiple tables as output rather than just the single table that the summary action produces when running without by groupings. The last table is called ByGroupInfo. This table contains information about all of the by grouping tables that follow. This information is more useful if there are a large number of by groupings and you want to know about them in advance so that you can prepare for the tables as they come from the server. The remaining tables correspond to each of the by groupings.

Each by grouping generates a new output table. The name of the key for that output is the name of the table prefixed by a label ByGroup#, where # is the index of the grouping (starting with 1). Without by grouping applied, the summary action creates a table called Summary. With by grouping applied, the result keys are ByGroup1.Summary, ByGroup2.Summary, and so on.

You can also set the groupby parameter as an option in the table argument, as shown in the following example.

```
> cas.simple.summary(
+   conn,
+   table = list(name = 'iris', groupby = 'Species'),
+   subset = c("min","max"))
```

```
$ByGroup1.Summary
        Column Min Max
1 Sepal.Length 4.3 5.8
2  Sepal.Width 2.3 4.4
3 Petal.Length 1.0 1.9
4  Petal.Width 0.1 0.6
```

```
$ByGroup2.Summary
        Column Min Max
1 Sepal.Length 4.9 7.0
2  Sepal.Width 2.0 3.4
3 Petal.Length 3.0 5.1
4  Petal.Width 1.0 1.8

$ByGroup3.Summary
        Column Min Max
1 Sepal.Length 4.9 7.9
2  Sepal.Width 2.2 3.8
3 Petal.Length 4.5 6.9
4  Petal.Width 1.4 2.5

$ByGroupInfo
     Species   Species_f      _key_
1      setosa      setosa setosa
2 versicolor versicolor versicolor
3  virginica  virginica virginica
```

Concatenating By Groups

Depending on your workflow, you may want to combine all of the by-grouping tables into a single data set. Fortunately, there is a function supplied by the SWAT package that can be used to combine by-group tables into a single casDataFrame: rbind.bygroups.

```
> grpsumm <- cas.simple.summary(grptbl, subset = c("min","max"))
> rbind.bygroups(grpsumm)
```

```
$Summary
      Species       Column Min Max
1      setosa Sepal.Length 4.3 5.8
2      setosa  Sepal.Width 2.3 4.4
3      setosa Petal.Length 1.0 1.9
4      setosa  Petal.Width 0.1 0.6
5  versicolor Sepal.Length 4.9 7.0
6  versicolor  Sepal.Width 2.0 3.4
7  versicolor Petal.Length 3.0 5.1
8  versicolor  Petal.Width 1.0 1.8
9   virginica Sepal.Length 4.9 7.9
10  virginica  Sepal.Width 2.2 3.8
11  virginica Petal.Length 4.5 6.9
12  virginica  Petal.Width 1.4 2.5
```

Selecting Result Keys by Table Name

Unlike the summary action, some CAS actions return multiple result table names, including the correlation action. We won't explain the correlation action in detail, but we use it in our example to select a particular table name from a result set. We still use our grptbl variable from the previous example as our starting point.

```
> grpcorr <- cas.simple.correlation(grptbl)
> grpcorr
```

```
$ByGroup1.CorrSimple
       Variable  N  Mean    Sum    StdDev Minimum Maximum
1 Sepal.Length 50 5.006 250.3 0.3524897     4.3     5.8
2  Sepal.Width 50 3.428 171.4 0.3790644     2.3     4.4
3 Petal.Length 50 1.462  73.1 0.1736640     1.0     1.9
4  Petal.Width 50 0.246  12.3 0.1053856     0.1     0.6

$ByGroup1.Correlation
       Variable Sepal.Length Sepal.Width Petal.Length Petal.Width
1 Sepal.Length    1.0000000   0.7425467    0.2671758   0.2780984
2  Sepal.Width    0.7425467   1.0000000    0.1777000   0.2327520
3 Petal.Length    0.2671758   0.1777000    1.0000000   0.3316300
4  Petal.Width    0.2780984   0.2327520    0.3316300   1.0000000

$ByGroup2.CorrSimple
       Variable  N  Mean    Sum    StdDev Minimum Maximum
1 Sepal.Length 50 5.936 296.8 0.5161711     4.9     7.0
2  Sepal.Width 50 2.770 138.5 0.3137983     2.0     3.4
3 Petal.Length 50 4.260 213.0 0.4699110     3.0     5.1
4  Petal.Width 50 1.326  66.3 0.1977527     1.0     1.8

$ByGroup2.Correlation
       Variable Sepal.Length Sepal.Width Petal.Length Petal.Width
1 Sepal.Length    1.0000000   0.5259107    0.7540490   0.5464611
2  Sepal.Width    0.5259107   1.0000000    0.5605221   0.6639987
3 Petal.Length    0.7540490   0.5605221    1.0000000   0.7866681
4  Petal.Width    0.5464611   0.6639987    0.7866681   1.0000000

$ByGroup3.CorrSimple
       Variable  N  Mean    Sum    StdDev Minimum Maximum
1 Sepal.Length 50 6.588 329.4 0.6358796     4.9     7.9
2  Sepal.Width 50 2.974 148.7 0.3224966     2.2     3.8
3 Petal.Length 50 5.552 277.6 0.5518947     4.5     6.9
4  Petal.Width 50 2.026 101.3 0.2746501     1.4     2.5

$ByGroup3.Correlation
       Variable Sepal.Length Sepal.Width Petal.Length Petal.Width
1 Sepal.Length    1.0000000   0.4572278    0.8642247   0.2811077
2  Sepal.Width    0.4572278   1.0000000    0.4010446   0.5377280
3 Petal.Length    0.8642247   0.4010446    1.0000000   0.3221082
4  Petal.Width    0.2811077   0.5377280    0.3221082   1.0000000

$ByGroupInfo
      Species  Species_f      _key_
1      setosa     setosa     setosa
2 versicolor versicolor versicolor
3  virginica  virginica  virginica
```

As you can see in the preceding output, the correlation action produces tables that are named CorrSimple and Correlation. These tables appear in each by grouping in the output. If we want to select only the Correlation tables from the output, we can again use the rbind.bygroups function to assist.

```
> result <- rbind.bygroups(grpcorr)
> result$Correlation
```

```
      Species       Variable Sepal.Length  Sepal.Width
1        setosa Sepal.Length    1.0000000    0.7425467
2        setosa  Sepal.Width    0.7425467    1.0000000
3        setosa Petal.Length    0.2671758    0.1777000
4        setosa  Petal.Width    0.2780984    0.2327520
5    versicolor Sepal.Length    1.0000000    0.5259107
6    versicolor  Sepal.Width    0.5259107    1.0000000
7    versicolor Petal.Length    0.7540490    0.5605221
8    versicolor  Petal.Width    0.5464611    0.6639987
9     virginica Sepal.Length    1.0000000    0.4572278
10    virginica  Sepal.Width    0.4572278    1.0000000
11    virginica Petal.Length    0.8642247    0.4010446
12    virginica  Petal.Width    0.2811077    0.5377280
   Petal.Length Petal.Width
1     0.2671758   0.2780984
2     0.1777000   0.2327520
3     1.0000000   0.3316300
4     0.3316300   1.0000000
5     0.7540490   0.5464611
6     0.5605221   0.6639987
7     1.0000000   0.7866681
8     0.7866681   1.0000000
9     0.8642247   0.2811077
10    0.4010446   0.5377280
11    1.0000000   0.3221082
12    0.3221082   1.0000000
```

The final topic for discussion about by groupings is the case in which a CAS action has support for multiple sets of by groupings.

Handling Multiple Sets of By Groups

Some actions, such as simple.mdSummary, enable you to specify multiple sets of by groups. The output for such actions is similar to what we saw in the examples in the preceding section, except that the result keys add another prefix: ByGroupSet#.

```
> grpmdsumm <- cas.simple.mdSummary(
+    tbl,
+    sets=list(
+      list(groupby='Sepal.Length'),
+      list(groupby='Petal.Length')
+      )
+    )
> names(grpmdsumm)
```

```
 [1] "ByGroupSet1.ByGroup1.MDSummary"  "ByGroupSet1.ByGroup10.MDSummary"
 [3] "ByGroupSet1.ByGroup11.MDSummary" "ByGroupSet1.ByGroup12.MDSummary"
  ...
```

```
[35] "ByGroupSet1.ByGroup9.MDSummary"   "ByGroupSet1.ByGroupInfo"
[37] "ByGroupSet2.ByGroup1.MDSummary"   "ByGroupSet2.ByGroup10.MDSummary"
[39] "ByGroupSet2.ByGroup11.MDSummary"  "ByGroupSet2.ByGroup12.MDSummary"
...
[79] "ByGroupSet2.ByGroup9.MDSummary"   "ByGroupSet2.ByGroupInfo"
```

As you can see from the preceding output, we get a ByGroupSet#.ByGroupInfo table for each by-group set followed by all of the tables for that by-group set.

Conclusion

If you are familiar with R data.frame, hopefully, the use of much of the API in CASTable objects helps you transition to the world of CAS. We covered the basics of using some of the supported data.frame functions, using various forms of indexing and iteration, filtering using expressions, creating computed columns, and using by groups. With all of that "under your belt," we can look at some more advanced statistical data in the following chapters.

Chapter 7: Data Exploration and Summary Statistics

Overview

The description of the columns in a table using tabular or visual output is typically the first step in a data analysis or a statistical modeling process. In this chapter, you will learn how to use CAS to explore and summarize data. Topics include summarizing continuous variables and categorical variables, data transformation, dimensional reduction, and related visualizations using built-in R functions and other packages.

Let's start by uploading the Organics Purchase data set from your local directory to CAS. In this example, conn is a CAS object that is connected to a session on a CAS server. For more information about managing your data in CAS, refer to Chapters 4–6. You can get the data for this book from the following Github repository: https://github.com/sassoftware/sas-viya-the-R-perspective/.

```
> out <- cas.table.loadTable(
+    conn,
+    path = 'yourSamplePath/organics.sashdat')
```

```
NOTE: Cloud Analytic Services made the HDFS file organics.sashdat available
as table ORGANICS in caslib yourSampleCaslib.
```

```
> organics <- defCasTable(conn,'organics')
```

Summarizing Continuous Variables

The simple action set provides several useful actions for univariate and multivariate data exploration and summarization. Similar to some action sets that you have learned about in previous chapters, such as builtins and table, the simple action set is also preloaded when you start a new CAS server. Therefore, you can use it on the organics variable, which is a CASTable reference to the data set organics.sashdat that you just uploaded.

Descriptive Statistics

Let's get started with a basic simple.summary action:

```
> cas.simple.summary(organics)
```

```
$Summary
            Column     Min      Max        N  NMiss        Mean          Sum
1           DemAffl    0.00    34.00  1606488  82460     8.711893     13995552
2           DemAge    18.00    79.00  1574340 114608    53.797152     84695008
3           PromTime   0.00    39.00  1667592  21356     6.564670     10947192
4    purchase_3mon   698.44  1188.06  1688948      0   950.027539   1604547112
5    purchase_6mon  1668.77  2370.87  1688948      0  2049.979250   3462308354
6    purchase_9mon  2624.09  3468.72  1688948      0  3070.016785   5185098708
7   purchase_12mon  3698.92  4684.88  1688948      0  4189.994798   7076683333
          Std      StdErr        Var          USS          CSS          CV
1    3.421045  0.002699106    11.70355  1.407294e+08     18801597  39.268672
2   13.205734  0.010524786   174.39140  4.830901e+09    274551188  24.547273
3    4.657008  0.003606302    21.68772  1.080310e+08     36166252  70.940468
4   50.067179  0.038525207  2506.72244  1.528598e+12   4233721342   5.270077
5   70.731010  0.054425412  5002.87581  7.106110e+12   8449592095   3.450328
6   86.588115  0.066626983  7497.50165  1.593100e+13  12662882920   2.820444
7  100.009042  0.076953988 10001.80855  2.966816e+13  16892524537   2.386854
       TValue ProbT     Skewness      Kurtosis
1    3227.695     0  0.8916212792   2.096090794
2    5111.472     0 -0.0798247122  -0.843977643
3    1820.333     0  2.2826359634   8.075535911
4   24659.894     0  0.0020064188  -0.003629311
5   37665.847     0  0.0017794945  -0.001335628
6   46077.680     0  0.0004033935   0.002077235
7   54448.053     0 -0.0025906213   0.002014442
```

In this example, you execute the summary action to generate descriptive statistics for all the numeric columns in the Organics data set. The default statistics that are generated by the summary action are described in the following table (7.1):

Table 7.1: Summary Statistics

Statistic	Description
Min	Minimum value
Max	Maximum value
N	Number of observations with nonmissing values
NMiss	Number of observations with missing values
Sum	Sum
Std	Standard Deviation
StdErr	Standard Error
Var	Variable
USS	Uncorrected Sum of Squares
CSS	Corrected Sum of Squares
CV	Coefficient of Variation
TValue	Value of *t*-statistics for hypothesis testing
ProbT	*p*-value of the *t*-statistics
Skewness	Skewness
Kurtosis	Kurtosis

Let's work on a specific set of columns. You can set the inputs parameter of the simple.summary action to obtain the descriptive statistics for specified columns:

```
> varlist <- c('DemAge', 'Purchase_12mon', 'Purchase_6mon')
> cas.simple.summary(organics, inputs = varlist)
```

```
$Summary
           Column     Min      Max       N   NMiss       Mean         Sum
1          DemAge    18.00    79.00 1574340  114608   53.79715    84695008
2 purchase_12mon  3698.92  4684.88 1688948       0 4189.99480  7076683333
3  purchase_6mon  1668.77  2370.87 1688948       0 2049.97925  3462308354
       Std      StdErr          Var          USS          CSS       CV
1  13.20573  0.01052479     174.3914 4.830901e+09    274551188 24.547273
2 100.00904  0.07695399  10001.8085 2.966816e+13  16892524537  2.386854
3  70.73101  0.05442541   5002.8758 7.106110e+12   8449592095  3.450328
    TValue ProbT     Skewness       Kurtosis
1  5111.472     0 -0.079824712 -0.843977643
2 54448.053     0 -0.002590621  0.002014442
3 37665.847     0  0.001779494 -0.001335628
```

This code generates summary statistics only for columns that you have specified: DemAge, Purchase_12mon, and Purchase_6mon. In the next example, we compute the summary statistics for the columns for recent purchase amounts and save the result in a local object.

```
> varlist <- c('Purchase_3mon', 'Purchase_6mon', 'Purchase_9mon',
+              'Purchase_12mon')
> result <- cas.simple.summary(organics, inputs = varlist)
```

The result is an R list object that contains a data.frame in the "Summary" key. A CAS action might return one or more tables referenced by the names of the list object. In this example, only one summary table has been returned by the simple.summary action.

```
> names(result)
```

```
[1] "Summary"
```

Result tables are returned in the form of casDataFrame objects. Many data.frame functions are implemented for casDataFrames as well. For example, you can use the t function to transform the output summary table.

```
> df <- result$Summary
> names(df)
```

```
[1]  "Column"    "Min"       "Max"       "N"         "NMiss"    "Mean"
[7]  "Sum"       "Std"       "StdErr"    "Var"       "USS"      "CSS"
[13] "CV"        "TValue"    "ProbT"     "Skewness"  "Kurtosis"
```

```
> df1 <- df[c('Column','Min','Mean','Max')]
> t(df1)
```

```
              1                2                3
Column  "purchase_3mon"  "purchase_6mon"  "purchase_9mon"
Min     "  698.44"       "1668.77"        "2624.09"
Mean    "  950.0275"     "2049.9792"      "3070.0168"
Max     "1188.06"        "2370.87"        "3468.72"
              4
Column  "purchase_12mon"
Min     "3698.92"
Mean    "4189.9948"
Max     "4684.88"
```

You can also use the built-in R function barplot to generate a bar chart (see Figure 7.1) using the summary statistics that are generated by the simple.summary action.

```
> barplot(df$Mean, names.arg = df$Column)
```

Figure 7.1: Barplot Using Result of simple.summary Action

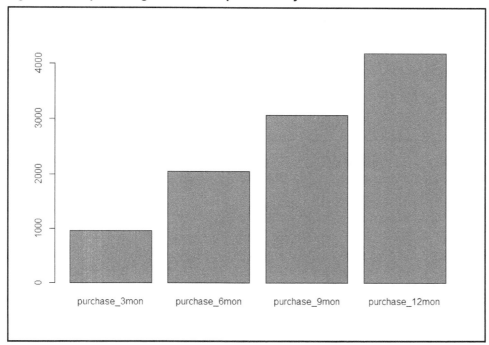

Sometimes it is useful to summarize your data by segments, which is accomplished with the groupby parameter. You can either set the groupby parameter as an attribute of the CASTable object or set it as an option in table argument of a CAS action.

```
> organics@groupby <- 'DemGender'
> result <- cas.simple.summary(organics, inputs = 'DemAge')
```

You can also use:

```
> result <- cas.simple.summary(
+    conn,
+    table = list(name = 'organics', groupby = 'DemGender'),
+    inputs = 'DemAge')
```

The groupby parameter groups the data into three segments for DemGender : F, M, or U. It also summarizes the column DemAge for each data segment, respectively. The summary statistics are returned in three different data frames.

```
> names(result)
```
```
[1] "ByGroup1.Summary" "ByGroup2.Summary" "ByGroup3.Summary" "ByGroupInfo"
```

In this case, the summary action also returns a table called ByGroupInfo that provides basic information about the by groups.

```
> result['ByGroupInfo']
$ByGroupInfo
```

	DemGender	DemGender_f	_key_
1	F	F	F
2	M	M	M
3	U	U	U

To merge these tables into one result table, you can use the rbind.bygroups function supplied by the SWAT package. It returns a new CASResults object with all by-group tables that are concatenated.

```
> result2 <- rbind.bygroups(result)
> result2['Summary']
```

	DemGender	DemGender_f	Column	Min	Max	N	NMiss	Mean	Sum
1	F	F	DemAge	18	79	861460	61864	52.88072	45554628
2	M	M	DemAge	18	79	411768	30172	54.53581	22456100
3	U	U	DemAge	18	79	301112	22572	55.40888	16684280

	Std	StdErr	Var	USS	CSS	CV	TValue
1	13.54665	0.01459534	183.5117	2567049492	158087808	25.61737	3623.123
2	12.81756	0.01997464	164.2897	1292310612	67649086	23.50301	2730.253
3	12.50470	0.02278815	156.3674	971541288	47083946	22.56803	2431.478

	ProbT	Skewness	Kurtosis
1	0	-0.01568211	-0.9278626
2	0	-0.10310664	-0.7631937
3	0	-0.18705608	-0.6433051

When you specify a group-by variable, the simple.summary action does not require any data shuffle to summarize data by segments. In fact, you can have as many group-by variables as you need. For example, the next action call summarizes DemAge by two columns: DemGender and DemHomeowner.

```
> result <- cas.simple.summary(
+    conn,
+    table = list(name = 'organics',
+                 groupby = c('DemGender','DemHomeowner')),
+    inputs = 'DemAge')

> names(result)
```

```
[1] "ByGroup1.Summary" "ByGroup2.Summary" "ByGroup3.Summary"
[4] "ByGroup4.Summary" "ByGroup5.Summary" "ByGroup6.Summary"
[7] "ByGroupInfo"
```

```
> result$ByGroupInfo
```

```
    DemGender DemGender_f DemHomeowner DemHomeowner_f  _key_
1         F           F           No             No   FNo
2         F           F           Yes            Yes   FYes
3         M           M           No             No   MNo
4         M           M           Yes            Yes   MYes
5         U           U           No             No   UNo
6         U           U           Yes            Yes   UYes
```

```
> result2 <- rbind.bygroups(result 'Summary')
> result2
```

```
$Summary
  DemGender DemHomeowner Column Min Max      N  NMiss      Mean
1         F           No DemAge  18  79 560471  40011  52.88348
2         F          Yes DemAge  18  79 300989  21853  52.87559
3         M           No DemAge  18  79 267238  19499  54.52526
4         M          Yes DemAge  18  79 144530  10673  54.55530
5         U           No DemAge  18  79 195894  14747  55.41474
6         U          Yes DemAge  18  79 105218   7825  55.39797
       Sum      Std      StdErr      Var        USS        CSS
1 29639658 13.54306 0.01809007 183.4146 1670246692 102798374
2 15914970 13.55334 0.02470422 183.6931  896802800  55289422
3 14571222 12.81461 0.02478885 164.2144  838383850  43884151
4  7884878 12.82301 0.03372959 164.4296  453926762  23764850
5 10855416 12.50437 0.02825212 156.3592  632179774  30629668
6  5828864 12.50536 0.03855238 156.3840  339361514  16454259
        CV   TValue ProbT   Skewness    Kurtosis
1 25.60925 2923.342     0 -0.01670682 -0.9261177
2 25.63252 2140.346     0 -0.01377669 -0.9310965
3 23.50216 2199.588     0 -0.10119272 -0.7639596
4 23.50461 1617.432     0 -0.10664614 -0.7617396
5 22.56505 1961.436     0 -0.18655416 -0.6434898
6 22.57368 1436.953     0 -0.18799294 -0.6429416
```

To remove the grouping variables from the CASTable object, you can simply set the groupby parameter to an empty string. After the grouping variables are removed, the summary action computes summary statistics for the entire data set.

```
> organics@groupby <- ''
> result <- cas.simple.summary(organics, inputs = 'DemAge')
> result
```

```
$Summary
  Column Min Max       N  NMiss     Mean      Sum      Std
1 DemAge  18  79 1574340 114608 53.79715 84695008 13.20573
       StdErr      Var        USS       CSS       CV   TValue ProbT
1 0.01052479 174.3914 4830901392 274551188 24.54727 5111.472     0
    Skewness   Kurtosis
1 -0.07982471 -0.8439776
```

Histograms

Histograms are another useful visualization tool for summarizing continuous data. A histogram is an estimate of the probability distribution of a continuous variable by *equal-distance binning*. The CAS dataPreprocess action set provides a histogram action. In the next example, we execute the dataPreprocess.histogram action for the DemAge column:

```
> loadActionSet(conn, 'dataPreprocess')
```

```
NOTE: Added action set 'dataPreprocess'.
NOTE: Information for action set 'dataPreprocess':
NOTE:     dataPreprocess
NOTE:         rustats - Computes robust univariate statistics, centralized
                        moments, quantiles, and frequency distribution
                        statistics
NOTE:         impute - Performs data matrix (variable) imputation
NOTE:         outlier - Performs outlier detection and treatment
NOTE:         binning - Performs unsupervised variable discretization
NOTE:         discretize - Performs supervised and unsupervised variable
                        discretization
NOTE:         catTrans - Groups and encodes categorical variables using
                        unsupervised and supervised grouping techniques
NOTE:         histogram - Generates histogram bins and simple bin-based
                        statistics for numeric variables
NOTE:         transform - Performs pipelined variable imputation, outlier
                        detection and treatment, functional transformation,
                        binning, and robust univariate statistics to
                   evaluate the quality of the transformation
NOTE:         kde - Computes kernel density estimation
NOTE:         highCardinality - Performs randomized cardinality estimation
```

```
> result <- cas.dataPreprocess.histogram(
    organics,
    reqpacks = list(list(nicebinning = FALSE, nbins = 10)),

    inputs = c('Purchase_3mon')
    )
> result['BinDetails']
```

```
$BinDetails
        Variable BinSetId BinId BinLowerBnd BinUpperBnd
1  purchase_3mon        1     1     698.440     747.402
2  purchase_3mon        1     2     747.402     796.364
3  purchase_3mon        1     3     796.364     845.326
4  purchase_3mon        1     4     845.326     894.288
5  purchase_3mon        1     5     894.288     943.250
6  purchase_3mon        1     6     943.250     992.212
7  purchase_3mon        1     7     992.212    1041.174
8  purchase_3mon        1     8    1041.174    1090.136
9  purchase_3mon        1     9    1090.136    1139.098
10 purchase_3mon        1    10    1139.098    1188.060
    BinWidth NInBin      Mean      Std     Min      Max
1     48.962     45  735.9204 11.62972  698.44   747.30
2     48.962   1802  783.9022 10.76720  747.72   796.36
```

```
3     48.962  28782   830.0947 12.08115   796.37   845.32
4     48.962 193828   875.7615 13.15566   845.33   894.28
5     48.962 529051   921.1567 13.78227   894.29   943.24
6     48.962 597872   966.3418 13.86998   943.25   992.21
7     48.962 279514  1011.6947 13.36085   992.22  1041.17
8     48.962  53719  1057.3838 12.42920  1041.18  1090.12
9     48.962   4214  1103.7459 11.36581  1090.14  1139.07
10    48.962    121  1151.3658 10.67709  1139.32  1188.06
     MidPoint        Percent
1     722.921  0.002664380
2     771.883  0.106693634
3     820.845  1.704137724
4     869.807 11.476256226
5     918.769 31.324291808
6     967.731 35.399076822
7    1016.693 16.549591817
8    1065.655  3.180618941
9    1114.617  0.249504425
10   1163.579  0.007164223
```

Each row of the BinDetails output contains the binning information for a specific bin, including the lower and upper bounds, the bin width, and summary statistics per bin, such as counts, averages, standard deviations, minimums, maximums, and midpoints. You can also use the barplot function to create a histogram, as shown in Figure 7.2.

```
> df = result$BinDetails
> barplot(df$Percent, names.arg = df$MidPoint, cex.names = 0.5,
        xlab='Purchase_3mon', ylab='Percent', col = '#1f77b4')
```

Figure 7.2: Histogram of Purchase_3mon with 10 Bins and nicebinning = FALSE

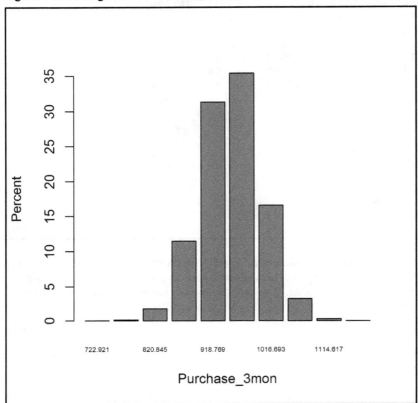

In the preceding example, we set nicebinning to FALSE, which is an option to request coarser upper bounds and lower bounds. Such a calculation usually returns nicer cutpoints for visualization and interpretation. A setting of TRUE for nicebinning generates the following barplot visualization, shown in Figure 7.3.

```
> result <- cas.dataPreprocess.histogram(
+    organics,
+    reqpacks = list(list(nicebinning = TRUE, nbins = 10)),

+    inputs = c('Purchase_3mon')
+ )
> df = result$BinDetails
> barplot(df$Percent, names.arg = df$MidPoint, cex.names = 0.5,
+         xlab='Purchase_3mon', ylab='Percent', col = '#1f77b4')
```

Figure 7.3: Histogram of Purchase_3mon with 10 Bins and nicebinning = TRUE

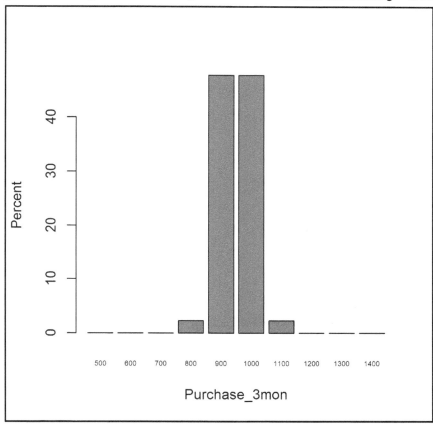

Compared to the previous graph shown in Figure 7.2, the nicebinning option allows bounds in Figure 7.3 only at a step size of 100, which is determined automatically by the range of the variable Purchase_3mon. Sometimes you need to increase the number of bins when you use the nicebinning option to get a better understanding of the data. In this example (see Figure 7.4), we change the number of bins from 10 to 25.

```
> result <- cas.dataPreprocess.histogram(
+    organics,
+    reqpacks = list(list(nicebinning = TRUE, nbins = 25)),
+    inputs = c('Purchase_3mon')
+ )
> df = result$BinDetails
> barplot(df$Percent, names.arg = df$MidPoint, cex.names = 0.5,
+         xlab='Purchase_3mon', ylab='Percent', col = '#1f77b4')
```

Figure 7.4: Histogram of Purchase_3mon with 25 Bins and nicebinning = TRUE

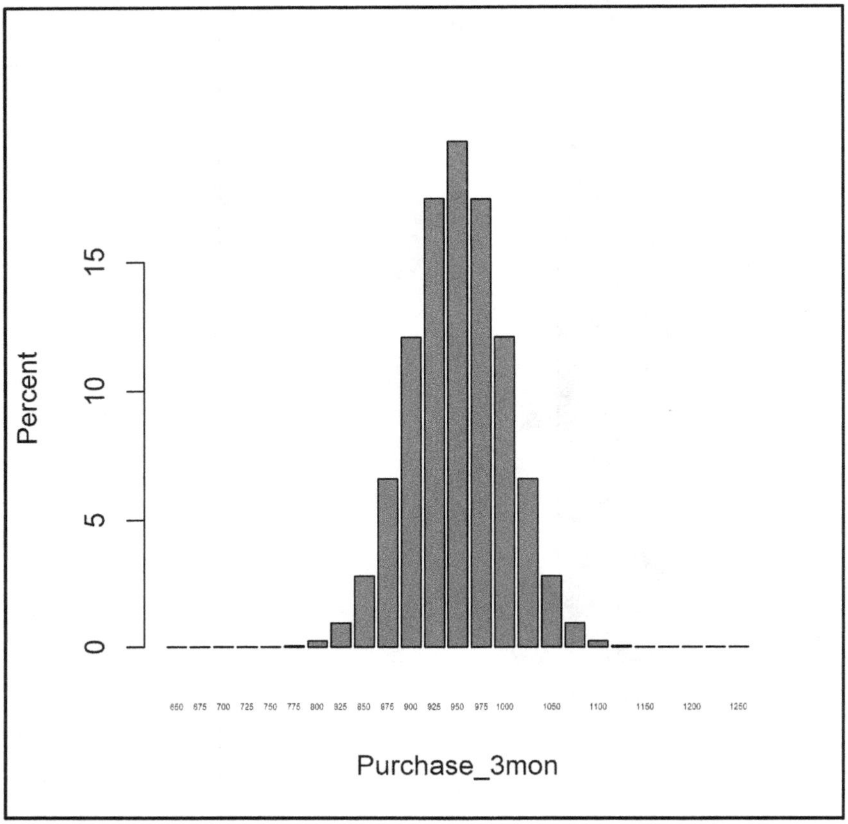

In this next example shown in Figure 7.5, 25 is chosen as the step size. The dataPreprocess.histogram action also enables you to set the bins manually. For example, you can force a fixed-bin width of 50. In this case, you don't need to specify a number of bins because it is determined automatically.

```
> result <- cas.dataPreprocess.histogram(
+     organics,
+     reqpacks = list(list(binwidth = 50)),

+     inputs = c('Purchase_3mon')
+ )
> df = result$BinDetails
> barplot(df$Percent, names.arg = df$MidPoint, cex.names = 0.5,
+           xlab='Purchase_3mon', ylab='Percent', col = '#1f77b4')
```

Figure 7.5: Histogram of Purchase_3mon with binwidth = 50

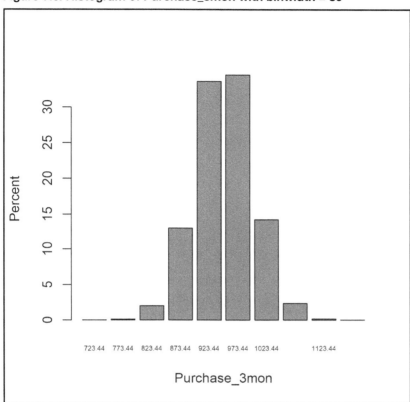

Similar to the simple.summary action, the dataPreprocess.histogram action also supports group-by processing. You can request a histogram for each data segment that is defined by a set of group-by columns.

```
> organics@groupby <- c('DemGender', 'DemAgeGroup')

> result <- cas.dataPreprocess.histogram(
+    organics,
+    reqpacks = list(list(nicebinning = TRUE, nbins = 20)),
+    inputs = c('DemAffl')

+ )
> names(result)
```

```
 [1] "ByGroup1.BinDetails"   "ByGroup10.BinDetails"
 [3] "ByGroup11.BinDetails"  "ByGroup12.BinDetails"
 [5] "ByGroup2.BinDetails"   "ByGroup3.BinDetails"
 [7] "ByGroup4.BinDetails"   "ByGroup5.BinDetails"
 [9] "ByGroup6.BinDetails"   "ByGroup7.BinDetails"
[11] "ByGroup8.BinDetails"   "ByGroup9.BinDetails"
[13] "ByGroupInfo"
```

Except for the ByGroupInfo table, each result table contains the histogram statistics for one by group. The ByGroupInfo table provides an overview of the by-group levels:

```
> result['ByGroupInfo']
$ByGroupInfo
   DemGender DemGender_f DemAgeGroup DemAgeGroup_f     _key_
1          F           F      middle        middle  Fmiddle
2          F           F      senior        senior  Fsenior
3          F           F     unknown       unknown Funknown
4          F           F       young         young   Fyoung
5          M           M      middle        middle  Mmiddle
6          M           M      senior        senior  Msenior
7          M           M     unknown       unknown Munknown
8          M           M       young         young   Myoung
9          U           U      middle        middle  Umiddle
10         U           U      senior        senior  Usenior
11         U           U     unknown       unknown Uunknown
12         U           U       young         young   Uyoung
```

Next, let's collect the statistics from the output table and use the barplot function to generate a grid of histograms (Figure 7.6). The all_df list contains the title of each histogram as well as its data. You must define the layout of the grid using the par function and then generate visualizations, one by one, using the barplot function.

```
> all_df = list()
> all_df[['Gender=Female, AgeGroup=Middle']]  = result$ByGroup1.BinDetails
> all_df[['Gender=Female, AgeGroup=Senior']]  = result$ByGroup1.BinDetails
> all_df[['Gender=Female, AgeGroup=Unknown']] = result$ByGroup1.BinDetails
> all_df[['Gender=Female, AgeGroup=Young']]   = result$ByGroup1.BinDetails
>
> par(mfrow=c(2,2))
>
> for (this_title in names(all_df)){
+   barplot(all_df[[this_title]]$Percent,
+           names.arg = all_df[[this_title]]$MidPoint,
+           xlab='DemAffl', ylab='Percent', col = '#1f77b4',
+           main = this_title)
+   }
```

Figure 7.6: Histogram with By Groups

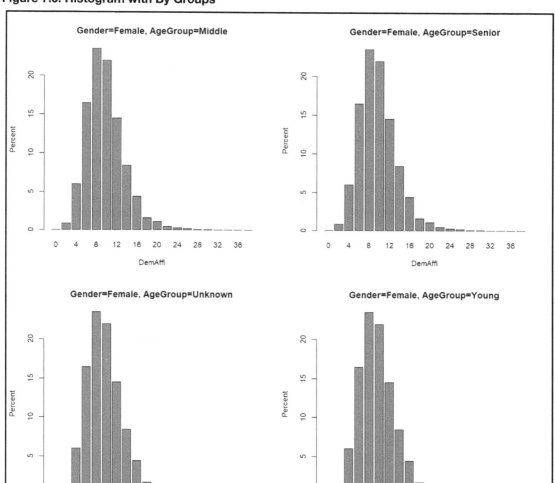

Percentiles

Percentile is another useful tool for numeric data exploration. A percentile is a statistic for indicating a variable that has a given percentage of observations that is below the percentile value. In CAS, such statistics are available in the percentile action set.

```
> cas.builtins.loadActionSet(conn, 'percentile')
```

```
NOTE: Added action set 'percentile'.
NOTE: Information for action set 'percentile':
NOTE:     percentile
NOTE:         percentile - Calculate quantiles and percentiles
```

```
NOTE:           boxPlot - Calculate quantiles, high and low whiskers, and
outliers
NOTE:           assess - Assess and compare models
```

```
> organics@groupby <- ''
> cas.percentile.percentile(organics, inputs = 'DemAge')
```

```
$Percentile
   Variable Pctl Value Converged
1    DemAge   25    44          1
2    DemAge   50    54          1
3    DemAge   75    64          1
```

By default, the percentile.percentile action computes the 25th, the 50th, and the 75th percentiles for the selected columns. The preceding result shows that 25% of the observations of DemAge are less than 44, and half of the observations of DemAge are less than 54. Finally, 75% of the observations of DemAge are less than 64. These three percentiles are also known as the first quartile (Q1), the second quartile (median), and the third quartile (Q3).

The last column of the output table indicates whether the percentile calculation has converged. The CAS percentile.percentile action does not sort the data to find the percentiles. Instead, it has a numeric approximation algorithm to estimate the percentile, and the last column indicates whether the algorithm has converged. Two parameters maxiters and epsilon are used to control the accuracy of the approximation.

You can also specify the percentage values directly using the values option. For example, you can get a list of percentiles for percentage from 5% to 90% with a step size of 5% as follows.

```
> result <- cas.percentile.percentile(
+    organics,
+    inputs = 'DemAge',
+    values = list(seq(5,90,5)))
> result
```

```
$Percentile
    Variable Pctl Value Converged
1     DemAge    5    32          1
2     DemAge   10    36          1
3     DemAge   15    39          1
4     DemAge   20    41          1
5     DemAge   25    44          1
6     DemAge   30    46          1
7     DemAge   35    48          1
8     DemAge   40    50          1
9     DemAge   45    52          1
10    DemAge   50    54          1
11    DemAge   55    56          1
12    DemAge   60    58          1
13    DemAge   65    60          1
14    DemAge   70    62          1
15    DemAge   75    64          1
16    DemAge   80    66          1
17    DemAge   85    69          1
18    DemAge   90    72          1
```

The percentile.percentile action supports group-by processing as well:

```
> organics@groupby <- 'DemGender'
> result <- cas.percentile.percentile(
+   organics,
+   inputs = 'DemAge',
+   values = list(seq(5,90,5)))
```

In this case, we can use the ggplot2 package to visually compare the distributions of DemAge across three different levels of the DemGender column, as shown in Figure 7.7.

```
> library('ggplot2')
> ggplot(result2, aes(Pctl, Value, shape = factor(DemGender))) +
+   geom_point(aes(colour = factor(DemGender)))
```

Figure 7.7: Compare the Distributions of DemAge across Three Different Levels of the DemGender

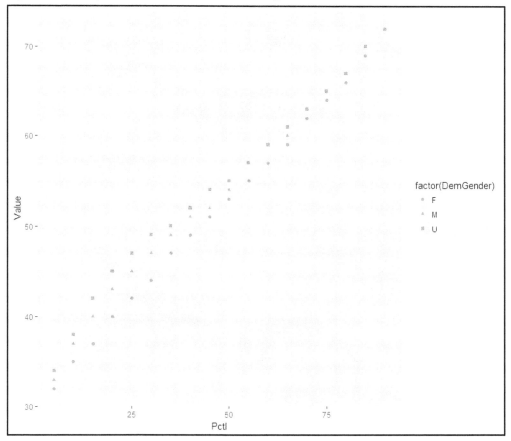

Correlations

The simple action set also provides a correlation action that generates a matrix of Pearson product-moment correlation coefficients for a set of variables.

```
> cas.simple.correlation(organics)
```

```
$CorrSimple
         Variable        N        Mean          Sum     StdDev Minimum Maximum
1          DemAffl 1606488    8.711893    13995552    3.421045    0.00   34.00
2           DemAge 1574340   53.797152    84695008   13.205734   18.00   79.00
3         PromTime 1667592    6.564670    10947192    4.657008    0.00   39.00
4    purchase_3mon 1688948  950.027539  1604547112   50.067179  698.44 1188.06
5    purchase_6mon 1688948 2049.979250  3462308354   70.731010 1668.77 2370.87
6    purchase_9mon 1688948 3070.016785  5185098708   86.588115 2624.09 3468.72
7   purchase_12mon 1688948 4189.994798  7076683333  100.009042 3698.92 4684.88


$Correlation
         Variable        DemAffl         DemAge       PromTime purchase_3mon
1          DemAffl  1.0000000000 -0.1377665101 -0.0297355863    0.001571352
2           DemAge -0.1377665101  1.0000000000  0.2072208849   -0.001160323
3         PromTime -0.0297355863  0.2072208849  1.0000000000   -0.000304101
4    purchase_3mon  0.0015713517 -0.0011603234 -0.0003041010    1.000000000
5    purchase_6mon  0.0016859128 -0.0010084839 -0.0005636005    0.707468622
6    purchase_9mon  0.0014587173 -0.0005407331 -0.0017200634    0.578366813
7   purchase_12mon  0.0009650401 -0.0005871620 -0.0022550974    0.500961133
   purchase_6mon purchase_9mon purchase_12mon    Nobs1    Nobs2    Nobs3
Nobs4
1  0.0016859128  0.0014587173   0.0009650401 1606488 1498264 1586728
1606488
2 -0.0010084839 -0.0005407331  -0.0005871620 1498264 1574340 1554656
1574340
3 -0.0005636005 -0.0017200634  -0.0022550974 1586728 1554656 1667592
1667592
4  0.7074686218  0.5783668132   0.5009611335 1606488 1574340 1667592
1688948
5  1.0000000000  0.8165998307   0.7073868090 1606488 1574340 1667592
1688948
6  0.8165998307  1.0000000000   0.8660580646 1606488 1574340 1667592
1688948
7  0.7073868090  0.8660580646   1.0000000000 1606488 1574340 1667592
1688948
      Nobs5    Nobs6    Nobs7
1 1606488 1606488 1606488
2 1574340 1574340 1574340
3 1667592 1667592 1667592
4 1688948 1688948 1688948
5 1688948 1688948 1688948
6 1688948 1688948 1688948
7 1688948 1688948 1688948
```

By default, the simple.correlation action uses all of the numeric columns and also computes the univariate statistics for each column as a by-product of computing the Pearson's correlation coefficients. The Nobs columns display the number of nonmissing observations for each pair of variables used to compute the correlation coefficients. These columns are added only when the number of observations used to compute the correlation coefficients differ due to the use of pairwise deletion. You can use the excludePairStats parameter to disable the display of the Nobs columns, but pairwise deletion is still used.

It also provides a simple flag to disable the output of the univariate statistics. To select a set of columns instead of all numeric columns, you must set the inputs option of the simple.correlation action.

```
> varlist <- c('DemAffl', 'DemAge', 'purchase_3mon')
> cas.simple.correlation(organics,inputs=varlist, simple=FALSE)
```

```
$Correlation
       Variable      DemAffl        DemAge purchase_3mon     Nobs1     Nobs2
1       DemAffl  1.000000000  -0.137766510    0.001571352  1606488  1498264
2        DemAge -0.137766510   1.000000000   -0.001160323  1498264  1574340
3 purchase_3mon  0.001571352  -0.001160323    1.000000000  1606488  1574340
      Nobs3
1 1606488
2 1574340
3 1688948
```

Summarizing Categorical Variables

In the previous section, we examined some useful actions in the simple, dataPreprocess, and percentile action sets of CAS for summarizing and exploring continuous variables. The simple action set also provides several actions for categorical data analysis, such as these actions: distinct, freq, topk, and crosstab.

Distinct Counts

Before you work on any categorical data, it is useful to first check the cardinality of the data using the simple.distinct action.

```
> organics <- defCasTable(conn,'organics')
> cas.simple.distinct(organics)
```

```
$Distinct
           Column NDistinct   NMiss Trunc
1              ID   1688948       0     0
2         DemAffl        34   82460     0
3          DemAge        63  114608     0
4       DemGender         3       0     0
5     DemHomeowner         2       0     0
6      DemAgeGroup         4       0     0
7      DemCluster        56       0     0
8          DemReg         6       0     0
9         DemTVReg        14       0     0
10        DemFlag1         2       0     0
11        DemFlag2         2       0     0
12        DemFlag3         2       0     0
13        DemFlag4         2       0     0
14        DemFlag5         2       0     0
15        DemFlag6         2       0     0
```

16	DemFlag7	2	0	0
17	DemFlag8	2	0	0
18	PromClass	4	0	0
19	PromTime	40	21356	0
20	TargetBuy	2	0	0
21	Bought_Beverages	2	0	0
22	Bought_Bakery	2	0	0
23	Bought_Canned	2	0	0
24	Bought_Dairy	2	0	0
25	Bought_Baking	2	0	0
26	Bought_Frozen	2	0	0
27	Bought_Meat	2	0	0
28	Bought_Fruits	2	0	0
29	Bought_Vegetables	2	0	0
30	Bought_Cleaners	2	0	0
31	Bought_PaperGoods	2	0	0
32	Bought_Others	2	0	0
33	purchase_3mon	32944	0	0
34	purchase_6mon	44997	0	0
35	purchase_9mon	54032	0	0
36	purchase_12mon	61444	0	0

By default, the simple.distinct action calculates two types of information for each column in the table: 1) the number of distinct values, and 2) the number of missing values for both character and numeric columns. Computing the exact distinct counts for the numeric columns and the unique identifiers (such as the ID column in the Organics table) could be slow and could require large memory footprints, but often the exact distinct counts for these columns are not useful because we need to know only whether some variables have distinct levels that exceed a certain threshold. The simple.distinct action provides the option maxnvals to skip the exact counting of the distinct levels for high cardinality variables.

```
> cas.simple.distinct(organics, maxnvals=500)
```

$Distinct

	Column	NDistinct	NMiss	Trunc
1	ID	500	0	1
2	DemAffl	34	82460	0
3	DemAge	63	114608	0
4	DemGender	3	0	0
5	DemHomeowner	2	0	0
6	DemAgeGroup	4	0	0
7	DemCluster	56	0	0
8	DemReg	6	0	0
9	DemTVReg	14	0	0
10	DemFlag1	2	0	0
11	DemFlag2	2	0	0
12	DemFlag3	2	0	0
13	DemFlag4	2	0	0
14	DemFlag5	2	0	0
15	DemFlag6	2	0	0
16	DemFlag7	2	0	0
17	DemFlag8	2	0	0
18	PromClass	4	0	0
19	PromTime	40	21356	0
20	TargetBuy	2	0	0

```
21    Bought_Beverages          2      0      0
22      Bought_Bakery           2      0      0
23      Bought_Canned           2      0      0
24       Bought_Dairy           2      0      0
25      Bought_Baking           2      0      0
26      Bought_Frozen           2      0      0
27        Bought_Meat           2      0      0
28      Bought_Fruits           2      0      0
29    Bought_Vegetables         2      0      0
30     Bought_Cleaners          2      0      0
31    Bought_PaperGoods         2      0      0
32       Bought_Others          2      0      0
33       purchase_3mon        500      0      1
34       purchase_6mon        500      0      1
35       purchase_9mon        500      0      1
36       purchase_12mon       500      0      1
```

In this example, the simple.distinct action loops through the data and has early stops for the distinct count calculations for these columns—ID, purchase_3mon, purchase_6mon, purchase_9mon, and purchase_12mon—after their current distinct counts reach 500. Therefore, in the output table, the number of distinct values for these columns are shown as 500. A new column Trunc is added as a binary indicator for whether the distinct count values are exact.

Note that in this data, the categorical variable with the highest cardinality is DemCluster, which has 56 distinct levels. Missing values is another useful item of information that you can access in the distinct action output. In this example, three columns—DemAffl, DemAge and PromTime—have a significant number of missing values.

In the preceding examples, we requested that the output be downloaded to the R client. You can also store the output of the distinct action in a table on the CAS server. You can then use the head function to download results from the output table.

```
> cas.simple.distinct(
+    organics,
+    maxnvals=500,
+    casout = list(name = 'distinctOutput'))
```

```
$OutputCasTables
                    casLib        Name Rows Columns
1    CASUSERHDFS(username) distinctOutput   36       4
```

```
> result <- defCasTable(conn, 'distinctOutput')
> head(result)
```

	Column	_NDis_	_NMiss_	_Truncated_
1	ID	0	0	1
2	DemAffl	34	82460	0
3	DemAge	63	114608	0
4	DemGender	3	0	0
5	DemHomeowner	2	0	0
6	DemAgeGroup	4	0	0

The table.columnInfo action generates basic information about each column, such as whether it is numeric (double, int32, or int64) or character (char or varchar). The distinct action further computes the cardinality of the columns. It is common to combine these two pieces of information to determine whether a column should be used as a categorical variable or a continuous variable. The following example merges the output tables from the table.columnInfo and simple.distinct actions using the cbind() function. Then it selects character columns with no more than 128 levels and numeric columns with no more than 16 levels as categorical variables. Also, it selects numeric columns with more than 16 levels as continuous variables. All other columns are dropped into the Others group.

```
> out1 <- cas.table.columnInfo(organics)$ColumnInfo
> out2 <- cas.simple.distinct(organics,maxnvals=1000)$Distinct
> out3 <- cbind(out1, out2)
>
> varname <- out3$Column
> vartype <- out3$Type
> varNDistinct <- out3$NDistinct
> catList <- c()
> contList <- c()
> otherList <- c()
>
> for (i in 1:length(varname)){
+   if (vartype[i] == 'char' & varNDistinct[i] <= 128)
+     catList <- c(catList, varname[i])
+   else if (vartype[i] == 'double' & varNDistinct[i] <= 16)
+     catList <- c(catList, varname[i])
+   else if (vartype[i] == 'double' & varNDistinct[i] > 16)
+     contList <- c(contList, varname[i])
+   else
+     otherList <- c(otherList, varname[i])
+ }
>
> varlist <- list(cats = catList,conts = contList, others = otherList)
> varlist
```

```
$cats
 [1] "DemGender"        "DemHomeowner"      "DemAgeGroup"
 [4] "DemCluster"       "DemReg"            "DemTVReg"
 [7] "DemFlag1"         "DemFlag2"          "DemFlag3"
[10] "DemFlag4"         "DemFlag5"          "DemFlag6"
[13] "DemFlag7"         "DemFlag8"          "PromClass"
[16] "TargetBuy"        "Bought_Beverages"  "Bought_Bakery"
[19] "Bought_Canned"    "Bought_Dairy"      "Bought_Baking"
[22] "Bought_Frozen"    "Bought_Meat"       "Bought_Fruits"
[25] "Bought_Vegetables" "Bought_Cleaners"  "Bought_PaperGoods"
[28] "Bought_Others"

$conts
[1] "DemAffl"        "DemAge"          "PromTime"        "purchase_3mon"
[5] "purchase_6mon"  "purchase_9mon"   "purchase_12mon"

$others
[1] "ID"
```

Frequency

The distinct action computes only the number of distinct levels of a categorical variable. To get the frequency distribution of a categorical variable, you must use the simple.freq action in the simple action set.

```
> cas.simple.freq(organics, inputs = "TargetBuy")
```

```
$Frequency
        Column      CharVar       FmtVar Level Frequency
1 TargetBuy Bought       Bought          1    418380
2 TargetBuy No           No              2   1270568
```

Note that this method is equivalent to using column indexing first and then calling the simple.freq action:

```
> cas.simple.freq(organics["TargetBuy"])
```

```
$Frequency
        Column      CharVar       FmtVar Level Frequency
1 TargetBuy Bought       Bought          1    418380
2 TargetBuy No           No              2   1270568
```

The output table of the simple.freq action contains one row for each level of the categorical variable, as shown in Table 7.2.

Table 7.2: Output Columns of simple.freq Action

Column Name	Column Label	Description
Column	Analysis Variable	Column name of the variable in the input data set
CharVar	Character Value	Unformatted value of the variable
FmtVar	Formatted Value	Formatted value of the variable (SAS formats)
Level	Level	Index of the distinct variable value
Frequency	Frequency	Frequency count

Next, let's visualize the result from the simple.freq action (see Figure 7.8).

```
> out <- cas.simple.freq(organics['TargetBuy'])
> df <- out$Frequency
> barplot(df$Frequency, names.arg = df$FmtVar,
+         xlab='TargetBuy', ylab='Frequency', col = '#1f77b4')
```

Figure 7.8: Frequency Distribution of TargetBuy

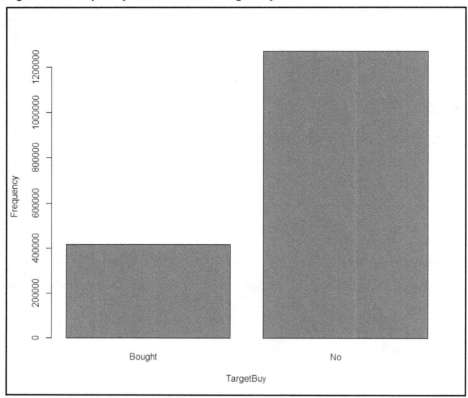

You can request multiple frequency tables in a single simple.freq call by adding more variables to the variable list of the CASTable object.

```
> cas.simple.freq(organics['TargetBuy','DemAgeGroup','DemHomeowner'])
```

```
$Frequency
        Column      CharVar      FmtVar  Level  Frequency
1  DemAgeGroup  middle        middle          1     991800
2  DemAgeGroup  senior        senior          2     532456
3  DemAgeGroup  unknown       unknown         3     114608
4  DemAgeGroup  young         young           4      50084
```

It is possible to use the simple.freq action on numeric variables as well. The next example depicted in Figure 7.9 uses the simple.freq action to generate a frequency table of the DemAge variable. In this example, the simple.freq action is equivalent to a dataPreprocess.histogram action with integer binning and a step size set to 1.

```
> out <- cas.simple.freq(organics['DemAge'], includemissing = FALSE)
> df <- out$Frequency
> barplot(df$Frequency, names.arg = df$NumVar,
+          xlab='Age', ylab='Frequency', col = '#1f77b4')
```

Figure 7.9: Frequency Distribution of Age

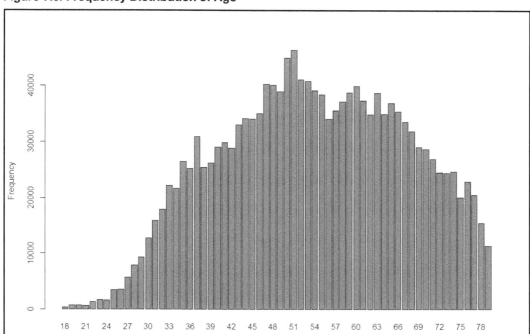

You can see from Figure 7.9 that the age of the customers in this data set has a mode of 51, and it is skewed to the left. The minimum age is 18 and the maximum age is 79. From the simple.distinct action, we also learn that this variable has a lot of missing values. So it is also helpful to enable the includemissing flag to display a missing bar on the bar chart, as shown in Figure 7.10.

```
> out <- cas.simple.freq(organics['DemAge'], includemissing = TRUE)
> df <- out$Frequency
> barplot(df$Frequency, names.arg = df$NumVar,
+          xlab='Age', ylab='Frequency', col = '#1f77b4')
```

Figure 7.10: Frequency Distribution of Age with includemissing = TRUE

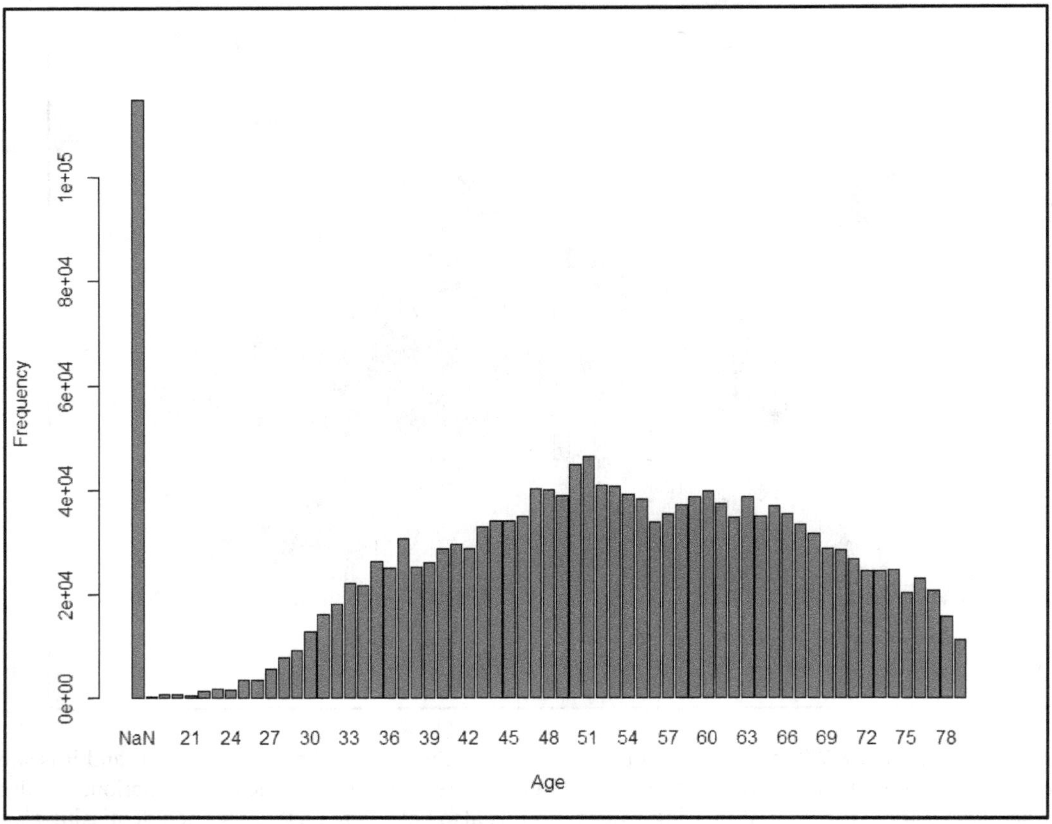

Top K

The simple.freq action returns a complete one-way frequency table. For high cardinality categorical variables, computation of the entire frequency tables might be time-consuming. In this case, the simple action set also provides a topk action to generate partial frequency tables.

```
> cas.simple.topK(organics['purchase_12mon'], topk=5, bottomk=0)
```

```
$Topk
          Column       FmtVar Rank
1 purchase_12mon       4684.88    1
2 purchase_12mon       4653.49    2
3 purchase_12mon       4646.95    3
4 purchase_12mon       4641.13    4
5 purchase_12mon       4631.91    5

$TopkMisc
          Column      N TruncatedTopk TruncatedBtmk ScoreOther
1 purchase_12mon  61444             0             0        NaN
```

Unlike the simple.freq action, the simple.topk action produces two output tables instead of one output table. The first table contains the top or bottom values of the input variables, and the second table contains a brief report for each input variable. You can loop through the list object to obtain the name of each table and its columns.

```
> result <- cas.simple.topK(organics['purchase_12mon'], topk=5,
bottomk=0)
> for (df in names(result)){
+    dfnames <- paste(names(result[[df]]),collapse=' ')
+    print(paste(df, ' table has: ', dfnames))
+ }
```

```
[1] "Topk    table has:  Column FmtVar Rank"
[1] "TopkMisc  table has:  Column N TruncatedTopk TruncatedBtmk ScoreOther"
```

Similar to the simple.freq action, the simple.topk action can be used to work on more than one variable and to compute both the bottom and top values of the variables. Simply specify the preferred variables when you subset the CASTable object, and specify the topk and bottomk parameters.

```
> cas.simple.topK(organics[c('purchase_12mon','DemAge')], topk=5,
bottomk=5)
```

```
$Topk
            Column       FmtVar   Rank
1   purchase_12mon      4684.88      1
2   purchase_12mon      4653.49      2
3   purchase_12mon      4646.95      3
4   purchase_12mon      4641.13      4
5   purchase_12mon      4631.91      5
6   purchase_12mon         3826  61444
7   purchase_12mon         3833  61443
8   purchase_12mon         3849  61442
9   purchase_12mon         3850  61441
10  purchase_12mon         3851  61440
11          DemAge           79      1
12          DemAge           78      2
13          DemAge           77      3
14          DemAge           76      4
15          DemAge           75      5
16          DemAge            .     63
17          DemAge           18     62
18          DemAge           19     61
19          DemAge           20     60
20          DemAge           21     59

$TopkMisc
            Column      N  TruncatedTopk  TruncatedBtmk  ScoreOther
1 purchase_12mon  61444              0              0         NaN
2          DemAge     63              0              0         NaN
```

The smallest reported value of DemAge is NaN. By default, the includemissing flag for the simple.topk action is enabled. Missing values are always treated as the lowest machine value by topk. This is the reason

NaN is reported as the bottom value of DemAge in the preceding example. To disable reporting NaN as the bottom value, set includemissing to FALSE.

The topk action is not only used to report the smallest and largest values of a variable, but also to report the levels of a categorical variable with the largest or smallest values with respect to another score variable. In the following example, the topk action is used to return the top and bottom TV regions with respect to customers' affluence score:

```
> cas.simple.topK(organics, inputs = 'DemTVReg',
+                 topk=3, bottomk=3,
+                 weight='DemAffl', agg='mean')
```

```
$Topk
      Column        FmtVar Rank     Score
1 DemTVReg Unknown            1  8.968397
2 DemTVReg East               2  8.921506
3 DemTVReg Yorkshire          3  8.752555
4 DemTVReg Ulster            14  8.493976
5 DemTVReg N East            13  8.512684
6 DemTVReg N West            12  8.533601


$TopkMisc
      Column  N TruncatedTopk TruncatedBtmk ScoreOther
1 DemTVReg 14              0             0   8.716037
```

The first result table shows that Unknown, East, and Yorkshire are the top three TV Regions with the highest average affluence scores 8.97, 8.92, and 8.75. TV Regions Ulster, N East, and N West are the bottom three TV Regions with the lowest average affluence scores 8.49, 8.51, and 8.53. The second table also indicates that there are 14 TV Regions in this data set, and the average affluence score for other TV regions (ScoreOther column) is 8.72.

You can also apply aggregators other than sample average in the simple.topk action. For example, let's get the TV regions with the highest and lowest total sales amounts.

```
> cas.simple.topK(organics, inputs = 'DemTVReg',
+                 topk=3, bottomk=3,
+                 weight='purchase_3mon', agg='sum')
```

```
$Topk
      Column        FmtVar Rank      Score
1 DemTVReg London             1  446844304
2 DemTVReg Midlands           2  225511094
3 DemTVReg S & S East         3  176541245
4 DemTVReg Border            14   14653126
5 DemTVReg Ulster            13   19199923
6 DemTVReg N Scot            12   23743227


$TopkMisc
      Column  N TruncatedTopk TruncatedBtmk ScoreOther
1 DemTVReg 14              0             0  698054193
```

The aggregators that are supported by the simple.topk action are max, mean, min, n, and sum.

Cross Tabulations

Crosstab is another useful action from the simple action set for categorical data exploration. This action requires a row variable and a col variable to generate a two-way frequency table.

```
> result <- cas.simple.crossTab(organics, row='DemAgeGroup',
col='DemGender')
> result
```

```
$Crosstab
  DemAgeGroup    Col1    Col2    Col3
1      middle  550240  258172  183388
2      senior  278312  143412  110732
3     unknown   61864   30172   22572
4       young   32908   10184    6992
```

The simple.crosstab output table has a DemAgeGroup column that indicates the levels of the row variable (DemAgeGroup) in the two-way frequency table. However, the levels of the column variable (DemGender) are not displayed in the output table because the simple.crosstab action stores the levels of the row variable in the labels instead of the names of the output columns:

```
> result$Crosstab@names
```

```
[1] "DemAgeGroup" "Col1"        "Col2"        "Col3"
```

```
> result$Crosstab@col.labels
```

```
[1] ""   "F" "M" "U"
```

In Chapter 3, we explained that the output tables of CAS actions are extensions of the data.frame object. Such extensions allow the output table to contain extra metadata that is contained in typical SAS data sets. Examples of metadata include the column label, the column format, and the data type. Such information is stored in attributes of the output data.frame. You can use the attributes function to retrieve that information.

In the preceding simple.crosstab example, the levels of the row variable are stored in the output column label instead of in the column names mainly because, in some situations, the levels might contain special characters that are not allowed in column names. You can retrieve the levels from the col.labels vector, and you can use the rename function to display the complete two-way frequency table:

```
> df <- result$Crosstab@df
> names(df)[-1] <- result$Crosstab@col.labels[-1]
> df
```

```
  DemAgeGroup       F       M       U
1      middle  550240  258172  183388
2      senior  278312  143412  110732
3     unknown   61864   30172   22572
4       young   32908   10184    6992
```

Very similar to what the SAS FREQ procedure provides, you can request chi-square tests of homogeneity and measures of associations.

```
> cas.simple.crossTab(organics, row='DemAgeGroup', col='DemGender',
+                     association=TRUE, chisq=TRUE)
```

```
$Association
                         Statistic       Value           ASE
1                            Gamma 0.022513030 1.227359e-03
2                 Kendall's Tau-B 0.012912614 7.057156e-04
3                 Stuart's Tau-c 0.011091773 6.061264e-04
4                                          NaN           NaN
5                 Somers' D C|R 0.013437504 7.345550e-04
6                 Somers' D R|C 0.012408227 6.780319e-04
7                                          NaN           NaN
8          Lambda Asymmetric C|R 0.000000000 0.000000e+00
9          Lambda Asymmetric R|C 0.000000000 0.000000e+00
10            Lambda Symmetric 0.000000000 0.000000e+00
11                                         NaN           NaN
12     Uncertainty Coefficient C|R 0.001286326 3.877649e-05
13     Uncertainty Coefficient R|C 0.001331922 4.011861e-05
14 Uncertainty Coefficient Symmetric 0.001308727 3.943370e-05
       LowerCL       UpperCL
1   0.020107452 0.024918609
2   0.011529437 0.014295791
3   0.009903787 0.012279758
4          NaN          NaN
5   0.011997802 0.014877205
6   0.011079309 0.013737145
7          NaN          NaN
8   0.000000000 0.000000000
9   0.000000000 0.000000000
10  0.000000000 0.000000000
11         NaN          NaN
12  0.001210325 0.001362327
13  0.001253291 0.001410553
14  0.001231439 0.001386016


$ChiSq
                    Statistic DF    Value Prob
1                  Chi-Square  6 4284.485    0
2 Likelihood Ratio Chi-Square  6 4334.500    0


$Crosstab
  DemAgeGroup   Col1   Col2   Col3
1      middle 550240 258172 183388
2      senior 278312 143412 110732
3     unknown  61864  30172  22572
4       young  32908  10184   6992
```

A two-way frequency table is usually displayed as a grouped or stacked bar chart. We can use the ggplot2 package to visualize the output from the crosstab action, as shown in Figure 7.11.

```
> result <- cas.simple.crossTab(organics,
+                               row='DemAgeGroup',
+                               col='DemGender')
> df <- result$Crosstab@df
>
> # use the reshape2 package to transform data from wide format
> # to long format
> library(reshape2)
> names(df)[-1] <- result$Crosstab@col.labels[-1]
> df_melt <- melt(df, id.vars = c('DemAgeGroup'))
> names(df_melt)[-1] = c('DemGender', 'Frequency')
>
> library(ggplot2)
> ggplot(df_melt, aes(DemGender, Frequency, fill = DemAgeGroup)) +
+   geom_bar(position = 'dodge', stat='identity') +
+   theme(legend.direction = 'vertical') +
+   scale_fill_brewer(palette = 'Set1')
```

Figure 7.11: Cross-Tabulation Result of DemGender and DemAgeGroup

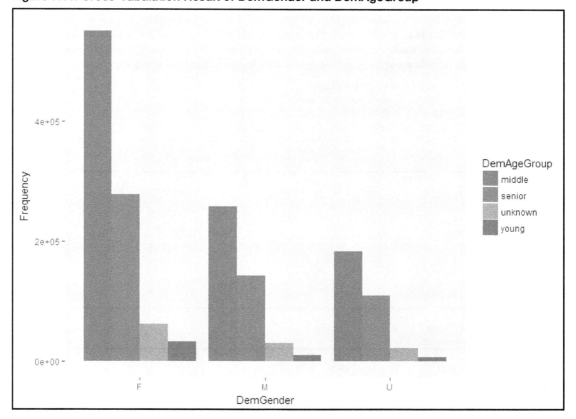

Note that we use the melt function from the reshape2 package to transform a data.frame from a wide format to a long format. You can also replace the dodge option with the stack option to generate a stacked bar chart instead of a grouped bar chart.

The crosstab action can do more than frequency counts. Similar to the simple.topk action, the simple.crosstab action can be used to specify an aggregation variable for each cell in the two-way table and to specify the type of aggregation. The following example shows how you can compute the total recent three-month sales amounts for the same two-way frequency table that is shown in Figure 7.11:

```
> organics <- defCasTable(conn,'organics')
> result <- cas.simple.crossTab(
+   organics, row='DemAgeGroup', col='DemGender',
+   weight='purchase_3mon', aggregators='sum')
> df <- result$Crosstab@df
> names(df)[-1] <- result$Crosstab@col.labels[-1]
> df
```

	DemAgeGroup	F	M	U
1	middle	522744714	245281125	174232600
2	senior	264361216	136232205	105216284
3	unknown	58781801	28665861	21439594
4	young	31276199	9675101	6640414

The results show that the female customers within the middle age group make the largest contribution in terms of purchase amount. Another difference between the simple.crosstab action and the SAS FREQ procedure is that the simple.crosstab action supports binning of numeric variables.

```
> result <- cas.simple.crossTab(
+   organics, row='DemAgeGroup', col='purchase_3mon',
+   colnbins=4, chisq=TRUE)
> df <- result$Crosstab@df
> names(df)[-1] <- result$Crosstab@col.labels[-1]
> df
```

	DemAgeGroup	(500, 700]	(700, 900]	(900, 1100]
1	middle	1	157358	833045
2	senior	0	84901	446820
3	unknown	0	18225	96230
4	young	0	7835	42188

	(1100, 1300]
1	1396
2	735
3	153
4	61

```
> result[['ChiSq']]
```

	Statistic	DF	Value	Prob
1	Chi-Square	9	6.223867	0.7173205
2	Likelihood Ratio Chi-Square	9	6.641251	0.6744137

In this case, the row variable DemAgeGroup is a categorical variable, and the column variable purchase_3mon is numeric. The simple.crosstab action uses equal-space binning to discretize the column variable purchase_3mon into four buckets before it computes the two-way frequency table. The *p*-values of the chi-square test show that these two variables are not correlated.

Similar to the SAS FREQ procedure, the simple.crosstab action also supports frequency tables with more than two dimensions. You must put the other dimensions into the group-by list. The next example returns a three-way frequency table as a collection of five different two-way frequency tables. For each table, you can also compute the chi-square likelihood of homogeneity.

```
> organics@groupby = 'DemReg'
> result <- cas.simple.crossTab(
+    organics, row='DemAgeGroup', col='DemGender', chisq=TRUE)
> for (table_name in names(result)){
+    df <- result[[table_name]]
+    if (grepl('ChiSq',table_name))
+      print(df[df$Statistic == 'Chi-Square',])
+ }
```

```
   Statistic DF    Value Prob
1 Chi-Square  6 2256.131    0
   Statistic DF    Value          Prob
1 Chi-Square  6 1096.161 1.412025e-233
   Statistic DF    Value          Prob
1 Chi-Square  6 146.5636  4.12041e-29
   Statistic DF    Value Prob
1 Chi-Square  6 2959.659    0
   Statistic DF    Value          Prob
1 Chi-Square  6 1137.124 1.935146e-242
   Statistic DF    Value          Prob
1 Chi-Square  6 913.5657 4.385563e-194
```

```
> # Reset the groupby parameter
> organics@groupby <- ''
```

Variable Transformation and Dimension Reduction

Once you get some ideas about a data set using the data exploration actions that were introduced in the previous two sections, it is often useful to transform some data columns because of "missing-ness" or non-linearity. The dataPreprocess action set provides a set of actions for variable imputation, variable binning, and dimension reduction.

```
> cas.builtins.loadActionSet(conn, 'dataPreprocess')
```

```
NOTE: Added action set 'dataPreprocess'.
NOTE: Information for action set 'dataPreprocess':
NOTE:    dataPreprocess
NOTE:        rustats - Computes robust univariate statistics, centralized
             moments, quantiles, and frequency distribution statistics
NOTE:        impute - Performs data matrix (variable) imputation
NOTE:        outlier - Performs outlier detection and treatment
NOTE:        binning - Performs unsupervised variable discretization
```

```
NOTE:        discretize - Performs supervised and unsupervised variable
             discretization
NOTE:        catTrans - Groups and encodes categorical variables using
             unsupervised and supervised grouping techniques
NOTE:        histogram - Generates histogram bins and simple bin-based
             statistics for numeric variables
NOTE:        transform - Performs pipelined variable imputation, outlier
             detection and treatment, functional transformation, binning,
and
             robust univariate statistics to evaluate the quality of the
             transformation
NOTE:        kde - Computes kernel density estimation
NOTE:        highCardinality - Performs randomized cardinality estimation
```

Variable Binning

We have shown the dataPreprocess.histogram action, which performs equal-spacing binning on one or multiple columns, and returns lower bounds, upper bounds, and frequencies for the bins for visualization purposes. The dataPreprocess.histogram action does not actually bin the data and store the binned results. This is accomplished by the binning action from the dataPreprocess action set.

```
> result <- cas.dataPreprocess.binning(
+    organics, inputs = 'purchase_3mon', tech='bucket',
+    casout = list(name = 'binnedData', replace = TRUE),
+    nBinsArray = 10)
> out_data <- defCasTable(conn, 'binnedData')
> head(out_data, n = 10L)
```

	BIN_purchase_3mon
1	5
2	4
3	6
4	4
5	6
6	5
7	6
8	5
9	5
10	5

In this example, we perform a bucket binning (which is also known as histogram binning or equal-space binning) on the purchase_3mon column with the number of bins equal to 10. You also must specify an output CAS table that contains the binned result. In most of the CAS actions, the option to specify a CAS table is casout. In the preceding example, the result table contains only one column that is the bin assignment (BIN_purchase_3mon).

The result object is a list object that is downloaded to the R client, and binnedData is actually a CAS table that is stored on the CAS server. If the CAS server is remote, the CAS table is not downloaded to the R client. You can always print out some observations from the casout table using the head function or the table.fetch action.

Unlike the dataPreprocess.histogram action, the dataPreprocess.binning action produces more than one result table, including VarTransInfo for information about the variable transformations, BinDetails for the actual binning information, and OutputCasTables for the casout table. The OutputCasTables table contains information about the CAS table, such as the table name, the library name, the number of observations, and the number of columns.

```
> result
```

```
$BinDetails
         Variable BinId BinLowerBnd BinUpperBnd BinWidth
1  purchase_3mon     1     698.440     747.402   48.962
2  purchase_3mon     2     747.402     796.364   48.962
3  purchase_3mon     3     796.364     845.326   48.962
4  purchase_3mon     4     845.326     894.288   48.962
5  purchase_3mon     5     894.288     943.250   48.962
6  purchase_3mon     6     943.250     992.212   48.962
7  purchase_3mon     7     992.212    1041.174   48.962
8  purchase_3mon     8    1041.174    1090.136   48.962
9  purchase_3mon     9    1090.136    1139.098   48.962
10 purchase_3mon    10    1139.098    1188.060   48.962
    NInBin      Mean       Std     Min      Max
1       45  735.9204  11.62972  698.44   747.30
2     1802  783.9022  10.76720  747.72   796.36
3    28782  830.0947  12.08115  796.37   845.32
4   193828  875.7615  13.15566  845.33   894.28
5   529051  921.1567  13.78227  894.29   943.24
6   597872  966.3418  13.86998  943.25   992.21
7   279514 1011.6947  13.36085  992.22  1041.17
8    53719 1057.3838  12.42920 1041.18  1090.12
9     4214 1103.7459  11.36581 1090.14  1139.07
10     121 1151.3658  10.67709 1139.32  1188.06

$OutputCasTables
                     casLib        Name    Rows Columns
1     CASUSERHDFS(username) binnedData 1688948       1

$VarTransInfo
       Variable           ResultVar NBins
1 purchase_3mon BIN_purchase_3mon    10
```

You can specify multiple variables in the dataPreprocess.binning action, and you can use a different number of bins for each column as well.

```
> result <- cas.dataPreprocess.binning(
+    organics,
+    inputs = c('purchase_3mon', 'purchase_6mon',
+               'purchase_9mon', 'purchase_12mon'),
+    tech='bucket',
+    casout = list(name = 'binnedData', replace = TRUE),
+    nBinsArray = c(4, 10, 20, 6))
> out_data <- defCasTable(conn, 'binnedData')
> head(out_data, n = 10L)
```

```
     BIN_purchase_12mon BIN_purchase_3mon
1                     3                  2
2                     3                  2
3                     2                  3
4                     4                  2
5                     3                  3
6                     4                  2
7                     4                  3
8                     5                  2
9                     3                  2
10                    3                  2
      BIN_purchase_6mon BIN_purchase_9mon
1                     7                 12
2                     6                 10
3                     4                  9
4                     6                 10
5                     7                 12
6                     6                 12
7                     6                 11
8                     7                 15
9                     6                 11
10                    5                  9
```

The dataPreprocess.binning action supports three types of binning techniques that are described in Table 7.3.

Table 7.3: Binning Techniques

Technique	Description
Bucket	Equal-width binning. Each bin has the same width.
Cutpts	Creates bins according to the user-specified cutpoints.
Quantile	Equal-frequency binning. Each bin has approximately the same number of observations.

The following example shows how to perform a quantile binning of four purchase amount columns. Also, it uses the copyallvars flag to pass all columns from the input table to the casout table.

```
> result <- cas.dataPreprocess.binning(
+    organics,
+    inputs = c('purchase_3mon', 'purchase_6mon',
+               'purchase_9mon', 'purchase_12mon'),
+    tech= 'Quantile',
+    casout = list(name = 'binnedData2', replace = TRUE),
+    copyallvars = TRUE,
+    nBinsArray = c(4, 4, 4, 4))
> out_data2 <- defCasTable(conn, 'binnedData2')
> names(out_data2)
```

```
 [1] "ID"           "DemAffl"
 [3] "DemAge"       "DemGender"
 [5] "DemHomeowner" "DemAgeGroup"
 [7] "DemCluster"   "DemReg"
 [9] "DemTVReg"     "DemFlag1"
[11] "DemFlag2"     "DemFlag3"
```

```
[13]  "DemFlag4"            "DemFlag5"
[15]  "DemFlag6"            "DemFlag7"
[17]  "DemFlag8"            "PromClass"
[19]  "PromTime"            "TargetBuy"
[21]  "Bought_Beverages"    "Bought_Bakery"
[23]  "Bought_Canned"       "Bought_Dairy"
[25]  "Bought_Baking"       "Bought_Frozen"
[27]  "Bought_Meat"         "Bought_Fruits"
[29]  "Bought_Vegetables"   "Bought_Cleaners"
[31]  "Bought_PaperGoods"   "Bought_Others"
[33]  "purchase_3mon"       "purchase_6mon"
[35]  "purchase_9mon"       "purchase_12mon"
[37]  "BIN_purchase_12mon"  "BIN_purchase_3mon"
[39]  "BIN_purchase_6mon"   "BIN_purchase_9mon"
```

You can get some idea about the binned data using the data exploration actions that were introduced previously. For example, you can generate a cross tabulation of two binned purchase amount columns.

```
> cas.simple.crossTab(out_data2, row='bin_purchase_3mon',
col='bin_purchase_12mon')
```

```
$Crosstab
  BIN_purchase_3mon   Col1    Col2    Col3    Col4
1                  1 203175  117851   70718   30483
2                  2 117668  124401  109169   70929
3                  3  71142  108972  124566  117614
4                  4  30231   70984  117812  203233
```

The dataPreprocess.binning action also provides a few options for cutpoints and observations assignment. The binning options are listed in Table 7.4.

Table 7.4: Binning Options

Binning Options	Description	Default Value
binmapping='left' | 'right'	Controls how to map values that fall at the boundary between consecutive bins. 'left' enables you to express the bins with [), [), ..., [] notation. 'right' enables [], (], ..., (] notation.	left
binmissing=TRUE | FALSE	When set to True, bins missing values are stored in a separate bin. The ID for this bin is 0.	TRUE
cutpts=*list*	Specifies the user-provided cutpoints for the cutpts binning technique.	Null
copyallvars=TRUE | FALSE	When set to True, all the variables from the input table are copied to the scored output table.	FALSE
copyvars=*list*	Specifies the names of the variables in the input table to use for identifying scored observations in the output table. The specified variables are copied to the output table.	Null
includeinputvars =TRUE | FALSE	When set to True, the analysis variables from the input table that are specified in the vars parameter are copied to the output table.	FALSE

Binning Options	Description	Default Value
pctldef=integer	Specifies the percentile definition to use. The definitions are numbered from 1 to 5. The default value, 0, specifies not to use a percentile definition.	0
pctlmaxniters=integer	Specifies the maximum number of iterations for percentile computation.	
pctltol=double	Specifies the tolerance for percentile computation.	1e-05

Another difference between the dataPreprocess.binning action and the dataPreprocess.histogram action is that the dataPreprocess.binning action can generate SAS DATA step code that represents the binning model. This code can be downloaded and saved for later use in scoring new observations. You can also apply the code to score a CAS table directly. Such DATA step code is often called score code when it is used to represent a data transformation or an analytical model.

The next example requests the score code for a bucket binning of the purchase_3mon variable, and then saves the score code into a local result table. Also, you can convert the score code that is stored in result['CodeGen'] into a text file.

```
> result <- cas.dataprePROCESS.binning(
+    organics,
+    inputs = 'purchase_3mon',
+    tech = 'bucket',
+    casout = list(name = 'binnedData2', replace = TRUE),
+    code = list(comment = TRUE, tabform = TRUE),
+    nBinsArray = 10)
> names(result)
```

```
[1] "BinDetails"      "CodeGen"         "OutputCasTables"
[4] "VarTransInfo"
```

```
> # score code is saved in this table
> # df = result['CodeGen']
> df = result$CodeGen
```

Variable Imputation

The impute action from the dataPreprocess action set manages missing values. You can use the simple.distinct action to check which columns have missing values.

```
> df <- cas.simple.distinct(organics)$Distinct
> df[df$NMiss > 0,]
```

```
    Column NDistinct  NMiss Trunc
2   DemAffl        34  82460     0
3    DemAge        63 114608     0
19 PromTime        40  21356     0
```

Let's run the dataPreprocess.impute action for the promotion time variable.

```
> cas.dataPreprocess.impute(organics, inputs = 'PromTime')
```

```
$ImputeInfo
  Variable ImputeTech    ResultVar        N NMiss
1 PromTime       Mean IMP_PromTime 1667592 21356
  ImputedValueContinuous
1               6.56467
```

The result table of the dataProcess.impute action contains information about the treatment of missing values for each specified variable. PromTime is a numeric variable and, by default, the dataPreprocess.impute action selects sample mean as the value to fill the observations with missing PromTime.

The preceding example only suggests the imputation method but it won't actually change the data. You must specify a casout table for the dataPreprocess.impute action in order to create a new data set with imputed values.

```
> cas.dataPreprocess.impute(
+    organics,
+    methodcontinuous = 'Median',
+    copyallvars = TRUE,
+    casout = list(name = 'cas.imputedData1', replace=TRUE),
+    inputs = 'PromTime')
```

```
$ImputeInfo
  Variable ImputeTech    ResultVar        N NMiss
1 PromTime     Median IMP_PromTime 1667592 21356
  ImputedValueContinuous
1              5

$OutputCasTables
                     casLib          Name  Rows Columns
1    CASUSERHDFS(username) cas.imputedData1 1688948      37
```

In this example, we add a casout table to the dataPreprocess.impute action, and request that all columns to be copied from the input table to the casout table. Also, the imputation technique is changed from mean to median. The dataPreprocess.impute action also enables you to specify the value to impute.

```
> cas.dataPreprocess.impute(
+    organics,
+    methodcontinuous = 'Value',
+    valuescontinuous = 0,
+    copyallvars = TRUE,
+    casout = list(name = 'cas.imputedData1', replace=TRUE),
+    inputs = 'PromTime')
```

```
$ImputeInfo
  Variable ImputeTech    ResultVar        N NMiss
1 PromTime      Value IMP_PromTime 1667592 21356
  ImputedValueContinuous
1                   0
```

```
$OutputCasTables
                     casLib            Name    Rows Columns
1     CASUSERHDFS(username) cas.imputedData1 1688948      37
```

The methodcontinuous option and the methodnominal option define the imputation treatment for continuous and categorical variables. The default imputations are mean and mode.

Table 7.5 lists techniques for continuous variables.

Table 7.5: Continuous Variable Techniques

Technique	Description
MAX	Replaces missing values with the maximum value.
MEAN	Replaces missing values with the mean.
MEDIAN	Replaces missing values with the median.
MIDRANGE	Replaces missing values with the mean of the maximum value and the minimum value.
MIN	Replaces missing values with the minimum value.
RANDOM	Replaces missing values with uniform random numbers.
VALUE	Replaces missing values with the values that are specified in the valuescontinuous parameters.

Table 7.6 lists techniques for categorical variables.

Table 7.6: Categorical Variable Techniques

Technique	Description
MODE	Replaces missing values with the mode.
VALUE	Replaces missing values with the values that are specified in the valuesnominal parameters.

Similar to other actions in the dataPreprocess action set, the impute action also generates score code that you can download so that you can repeat the imputation for a new data source.

```
> result <- cas.dataPreprocess.impute(
  organics,
  methodcontinuous = 'Value',
  valuescontinuous = 0,
  copyallvars = TRUE,
  casout = list(name = 'cas.imputedData1', replace=TRUE),
  code = list(comment = TRUE, tabform = TRUE),
  inputs = 'PromTime')
> result$CodeGen
```

```
                                   SASCode
1    /*------------------------------------------------------------------------
2    SAS Code Generated by Cloud Analytic Server for Impute Action
3    Date : 07Mar2016:11:12:45
4    Number of variables : 1
5    ------------------------------------------------------------------------*/
6    _igby_ = 0;
7    _tnn_ntrans_ = 1;
8
9    _fuzcmp_ = 0.000000000100;
10
11   array _tnn_vnames_{1} imp_PromTime ;
12
13   array _vnn_names_{1} PromTime ;
14
15   array _tnn_ntransvars_{1} _temporary_ (1 );
16
17   array _tv_nn_indices_{1} _temporary_ (1 );
18
19   imp_PromTime = .;
20
21   array _tnn_imputetype_{1} _temporary_ (2 );
22
23
24   array _tnn_imputeuniquevals_{1} _temporary_ (5 );
25
26
27   /*---------Iterate and score----------------*/
28
29   /*---------Count variables----------------*/
30   _ct_ = 0;
31   _impct_ = 0;
32
33   do _i_ = 1 to _tnn_ntrans_;
34   do _j_ = 1 to _tnn_ntransvars_{_i_};
35
36   if (_tnn_imputetype_{_i_} ~= 0) then
37   _impct_ + 1;
38   _ct_ + 1;
39   _numval_ = _vnn_names_{_tv_nn_indices_{_ct_}};
40
41   /*-------Apply Imputation phase--------------*/
42   if missing(_numval_) then
43   do;
44   if _tnn_imputetype_{_i_} = 0 then
45   goto _impute_done1_;
46   else _numval_ = _tnn_imputeuniquevals_{1 *_igby_ + _impct_};
47   end;
```

```
                                       SASCode
48  _impute_done1_:;
49  _tnn_vnames_{_ct_} = _numval_;
50  end;
51  end;
52
53       drop _igby_ _tnn_ntrans_ _fuzcmp_ _ct_ _impct_ _i_ _j_ _numval_ ;
```

Conclusion

In this chapter, we introduced actions in the simple and dataPreprocess action sets for data exploration, data summarization, and basic variable transformation. These actions enable you to learn about your data using different techniques to summarize continuous and categorical variables. You can also apply variable transformations, such as binning and imputation to reshape your data. Now that you understand your data, let's move on to Chapters 8 and 9 to learn about building analytic models.

Chapter 8: Modeling Continuous Variables

Overview

In this chapter, we explore several commonly used predictive models for a continuous dependent variable, including linear regressions, generalized linear models, and regression trees. The data set that is used in this chapter is the Cars data set that is available in the sas-viya-programming repository of the sassoftware account on GitHub. Upload the CSV data set directly from GitHub using the cas.read.csv function, as follows:

```
> cars <- cas.read.csv(conn, file =
'https://raw.githubusercontent.com/sassoftware/sas-viya-
programming/master/data/cars.csv')
> cas.table.tableInfo(cars)
```

```
$TableInfo
  Name Rows Columns IndexedColumns Encoding
1 CARS  428      15              0    utf-8
  CreateTimeFormatted    ModTimeFormatted
1  04Dec2017:19:58:01 04Dec2017:19:58:01
  AccessTimeFormatted JavaCharSet CreateTime     ModTime
1  04Dec2017:19:58:01        UTF8 1828036682 1828036682
  AccessTime Global Repeated View SourceName
1 1828036682      0        0    0
  SourceCaslib Compressed Creator Modifier
1                       0 username
```

```
> cas.table.columnInfo(cars)
```

```
$ColumnInfo
      Column ID   Type RawLength FormattedLength NFL NFD
1       Make  1 varchar        13              13   0   0
2      Model  2 varchar        39              39   0   0
3       Type  3 varchar         6               6   0   0
```

```
 4        Origin  4 varchar        6              6  0  0
 5     DriveTrain  5 varchar        5              5  0  0
 6          MSRP  6  double        8             12  0  0
 7       Invoice  7  double        8             12  0  0
 8     EngineSize  8  double        8             12  0  0
 9     Cylinders  9  double        8             12  0  0
10     Horsepower 10  double        8             12  0  0
11      MPG_City 11  double        8             12  0  0
12    MPG_Highway 12  double        8             12  0  0
13        Weight 13  double        8             12  0  0
14      Wheelbase 14  double        8             12  0  0
15        Length 15  double        8             12  0  0
```

Linear Regression

Linear regression is one of the most widely used statistical models for predictive modeling. The basic idea of a predictive model is to establish a function $y = f(x_1, x_2, \cdots, x_K)$ to predict the value of the dependent variable y that is based on the values of the predictors $X_1, X_2, \cdots, X_K$. Linear regression assumes that the function f is a linear combination of the predictors and an error term ε.

$$y = a + b_1 x_1 + b_2 x_2 + \cdots + b_K x_K + \varepsilon$$

Usually, we assume that ε follows a normal distribution with mean zero and variance σ^2. The parameters to be estimated in a linear model include the intercept α, the slopes $b_1, b_2, \cdots, b_K$, and the variance of the error term σ^2. These parameters are estimated using the least squares method.

The regression action set provides a glm action, which fits linear regression models using the least squares method. You must load the action set before you use the regression.glm action.

```
> cas.builtins.loadActionSet(conn, 'regression')
```

```
NOTE: Added action set 'regression'.
NOTE: Information for action set 'regression':
NOTE:      regression
NOTE:         glm - Fits linear regression models using the method of least
              squares
NOTE:         genmod - Fits generalized linear regression models
NOTE:         logistic - Fits logistic regression models
NOTE:         genmodScore - creates a table on the server that contains
              results from scoring observations by using a fitted model.
NOTE:         logisticScore - creates a table on the server that contains
              results from scoring observations by using a fitted model.
NOTE:         glmScore - creates a table on the server that contains results
              from scoring observations by using a fitted model.
```

Let's build a simple regression model using the Cars data to predict MSRP using the city miles per gallon (MPG) of the cars:

```
> cas.regression.glm(cars, target = 'MSRP', inputs = c('MPG_City'))
```

```
$ANOVA
  RowId       Source  DF          SS          MS    FValue
1 MODEL        Model   1  36380901274  36380901274  124.1344
```

```
2 ERROR             Error 426 124850717429   293076801      NaN
3 TOTAL Corrected Total 427 161231618703         NaN      NaN
          ProbF
1 1.783404e-25
2         NaN
3         NaN


$Dimensions
      RowId        Description Value
1 NEFFECTS    Number of Effects      2
2   NPARMS Number of Parameters      2

$FitStatistics
      RowId Description       Value
1      RMSE    Root MSE 1.711949e+04
2   RSQUARE    R-Square 2.256437e-01
3    ADJRSQ    Adj R-Sq 2.238260e-01
4       AIC         AIC 8.776260e+03
5      AICC        AICC 8.776316e+03
6       SBC         SBC 8.354378e+03
7 TRAIN_ASE         ASE 2.917073e+08

$ModelInfo
         RowId         Description Value
1          DATA         Data Source  CARS
2 RESPONSEVAR Response Variable  MSRP

$NObs
  RowId                     Description Value
1 NREAD Number of Observations Read    428
2 NUSED Number of Observations Used    428

$ParameterEstimates
      Effect Parameter DF  Estimate     StdErr    tValue
1 Intercept Intercept  1 68124.607 3278.9191  20.77654
2  MPG_City  MPG_City  1 -1762.135  158.1588 -11.14156
          Probt
1 1.006169e-66
2 1.783404e-25

$Timing
          RowId                Task       Time      RelTime
1          SETUP  Setup and Parsing 0.105124235 0.208397474
2   LEVELIZATION        Levelization 0.070063829 0.138893996
3 INITIALIZATION Model Initialization 0.258576155 0.512599378
4           SSCP    SSCP Computation 0.056274891 0.111558910
5        FITTING      Model Fitting 0.004421949 0.008766038
6        CLEANUP            Cleanup 0.005755186 0.011409037
7          TOTAL              Total 0.504441023 1.000000000
```

The ParameterEstimates table contains the estimation of parameters for the linear regression model. In the preceding example, the model returned by the glm action is shown as follows:

$$MSRP = 68124.606698 + -1762.135298 \times MPG\_CITY$$

The result tables also contain useful information about the model definition and model fitting. Table 8.1 summarizes the result tables.

Table 8.1: Summary of Result Tables

Table Name	Description
NObs	The number of observations that are read and used. Missing values are excluded, by default.
Dimension	The dimension of the model, including the number of effects and the number of parameters.
ANOVA	The Analysis of Variance table that measures the overall model fitting.
FitStatistics	The fit statistics of the model, such as R-square and root mean square error.
ParameterEstimates	The estimation of the regression parameters.
Timing	A timing of the subtasks of the glm action call.

To display only the parameter estimation table, you specify the names of the output table in the display option and rerun the linear1 model:

```
> cas.regression.glm(cars, target = 'MSRP', inputs = c('MPG_City'),
+                     display = list(names = "ParameterEstimates"))
```

```
$ParameterEstimates
      Effect Parameter DF   Estimate      StdErr     tValue
1 Intercept Intercept  1  68124.607  3278.9191   20.77654
2  MPG_City  MPG_City  1  -1762.135   158.1588  -11.14156
         Probt
1 1.006169e-66
2 1.783404e-25
```

So far, we have only used the regression.glm action example to estimate the parameters of the linear regression model. We haven't used the model to predict MSRP values of the cars. For prediction, you must specify an output table using the output option. You can also set the display.names option in order to display the output table:

```
> cas.regression.glm(
+    cars,
+    target = 'MSRP',
+    inputs = c('MPG_City'),
+    display = list(names = 'OutputCasTables'),
+    output = list(casOut = list(name='MSRPPrediction', replace = TRUE),
+                  copyvars = 'all'))
```

```
$OutputCasTables
                casLib              Name Label Rows Columns
1 CASUSERHDFS(username) MSRPPrediction         428      16
```

In the preceding example, when you submit the code, the regression.glm action first fits a linear regression model, and then it uses the fitted model to score the input data. Also, it creates the new table MSRPPrediction that contains the predicted MSRP values. The copyvars='all' option requests that the regression.glm action copy all columns from the CARS table to the MSRPPrediction table.

In the preceding example, the output column name for predicted MSRP is not specified. The regression.glm action automatically chooses Pred as the name.

You can summarize the predicted values using the summary action from the simple action set.

```
> cas.simple.summary(result1['Pred'])
```

```
$Summary
  Column        Min       Max   N NMiss      Mean        Sum
1    Pred -37603.51  50503.25 428       0  32774.86  14027638

         Std    StdErr        Var          USS          CSS       CV
1   9230.448  446.1706   85201174  496134704683  36380901274  28.1632

     TValue             ProbT  Skewness  Kurtosis
1  73.45813  2.182203e-244  -2.782072  15.79115
```

The regression.glm action can generate additional columns besides the predicted values. Table 8.2 summarizes the statistical outputs that the regression.glm action can generate.

Table 8.2: Summary of the Available Statistical Outputs of the regression.glm Action

Option	Description
pred	The predicted value. If you do not specify any output statistics, the predicted value is named Pred, by default.
resid	The residual, which is calculated as ACTUAL minus PREDICTED.
cooksd	The Cook's D influence statistic.
covratio	The standard influence of the observation on covariance of betas. The COVRATIO statistic measures the change in the determinant of the covariance matrix of the estimates by deleting the current observation.
dffits	The scaled measure of the change in the predicted value for the ith observation and is calculated by deleting the ith observation. A large value indicates that the observation is very influential in its neighborhood of the X space.
h	The leverage of the observation.
lcl	The lower bound of a confidence interval for an individual prediction.
ucl	The upper bound of a confidence interval for an individual prediction.
lclm	The lower bound of a confidence interval for the expected value of the dependent variable.
uclm	The upper bound of a confidence interval for the expected value of the dependent variable.
likedist	The likelihood displacement.
press	The ith residual divided by $1 - h$, where h is the leverage, and where the model has been refit without the ith observation
rstudent	The studentized residual with the current observation deleted.
stdi	The standard error of the individual predicted value.
stdp	The standard error of the mean predicted value.
stdr	The standard error of the residual.
student	The studentized residuals, which are the residuals divided by their standard errors.

The following example adds the residual values (observed MSRP values minus the predicted MSRP values) and the confidence intervals of the prediction to the output table:

```
> cas.regression.glm(
+    cars, target = 'MSRP', inputs = c('MPG_City'),
+    display = list(names = 'OutputCasTables'),
+    output = list(casOut = list(name='MSRPPrediction2', replace = TRUE),
+                  copyvars = 'all',
+                  pred  = 'Predicted_MSRP',
+                  resid = 'Residual_MSRP',
+                  lcl = 'LCL_MSRP',
+                  ucl = 'UCL_MSRP'
+                  ))
```

```
$OutputCasTables
               casLib          Name Label Rows Columns
1 CASUSERHDFS(username) MSRPPrediction2        428      19
```

```
> result2 <- defCasTable(conn, 'MSRPPrediction2')
```

The output table MSRPPredcition2 is a table that is saved on the CAS server. You have several ways to fetch or download a table from the CAS server. Since the Cars data set is relatively small, you can pull all observations from MSRPPredcition2 directly to the R client using the table.fetch action. Then you can use a visualization package in R, such as ggplot2 to observe and understand the model outcome.

```
> out1 <- cas.table.fetch(result2, to = 1000)$Fetch
> library(ggplot2)
> ggplot(out1, aes(Predicted_MSRP, Residual_MSRP,
+                  colour = Origin, shape = Origin)) +
+ geom_point()
```

Figure 8.1 shows a scatter plot of the predicted MSRP values and residuals.

Figure 8.1: Scatter Plot of the Predicted MSRP Values and Residuals

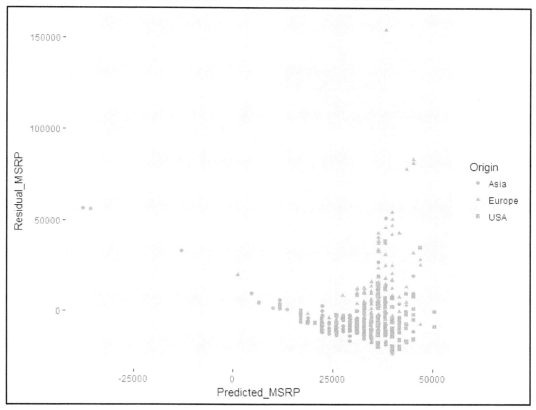

You can see that the model predicted large negative MSRP values for three observations. Let's print out these observations to find out what happens:

```
> result2[result2$Predicted_MSRP<0,
+          c('Predicted_MSRP','MSRP','MPG_City','Make','Model')]
```

```
  Predicted_MSRP  MSRP MPG_City    Make
1      -12933.62 20140       46   Honda
2      -37603.51 19110       60   Honda
3      -35841.38 20510       59 Toyota
                                  Model
1 Civic Hybrid 4dr manual (gas/electric)
2            Insight 2dr (gas/electric)
3              Prius 4dr (gas/electric)
```

All of these cars are fuel efficient with relatively high MPG_City. If you generate a scatter plot of the dependent variable MSRP and the predictor variable MPG_City, you can see that the data has some extreme outliers with high MSRP values or high city MPG that might not fit into the linear relationship assumption between these two variables.

```
> ggplot(out1, aes(MPG_City, MSRP,
+                      colour = Origin, shape = Origin)) +
+    geom_point()
```

Figure 8.2 shows a scatter plot of MPG_City and MSRP.

Figure 8.2: Scatter Plot of MPG_City and MSRP

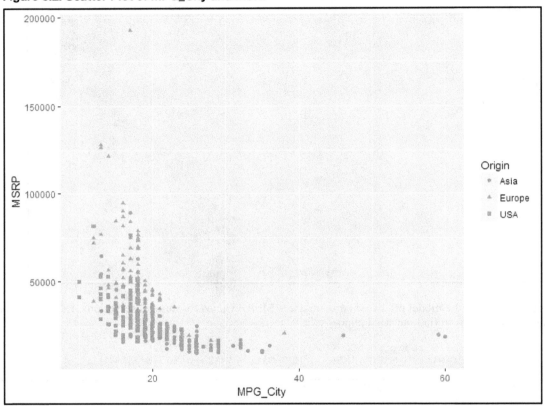

Outliers are observed when MSRP values are higher than $100,000 or MPG City is higher than 40.

Linear regression models are sensitive to outliers. There are several treatments for outliers to improve the prediction accuracy of a linear regression model. The simplest approach is to remove the outliers. In the next example, let's add a filter to remove cars with MPG_City that is greater than 40 or MSRP that is greater than 100,000. The R-square of the model is actually improved from 0.22 to 0.34.

```
> cars@where <- 'MSRP < 100000 and MPG_City < 40'
> cas.regression.glm(
+    cars, target = 'MSRP', inputs = c('MPG_City'),
+    display = list(names = 'FitStatistics'),
+    output = list(casOut = list(name='MSRPPrediction2', replace = TRUE),
+                      copyvars = 'all',
```

```
+                    pred  = 'Predicted_MSRP',
+                    resid = 'Residual_MSRP',
+                    lcl = 'LCL_MSRP',
+                    ucl = 'UCL_MSRP'
+    ))
```

```
$FitStatistics
       RowId Description        Value
1       RMSE    Root MSE 1.308617e+04
2     RSQUARE    R-Square 3.415360e-01
3     ADJRSQ    Adj R-Sq 3.399645e-01
4       AIC         AIC 8.406575e+03
5      AICC        AICC 8.406633e+03
6       SBC         SBC 7.991661e+03
7 TRAIN_ASE         ASE 1.704343e+08
```

You can see that we have a better residual plot after the outliers are removed from the model.

```
> result2 <- defCasTable(conn, 'MSRPPrediction2')
> out2 <- cas.table.fetch(result2, to = 1000)$Fetch
> ggplot(out2, aes(Predicted_MSRP, Residual_MSRP,
+                   colour = Origin, shape = Origin)) +
+    geom_point()
```

Figure 8.3 shows a scatter plot of predicted MSRP values and residuals, after excluding outliers.

Figure 8.3: Scatter Plot of Predicted MSRP Values and Residuals, after Excluding Outliers

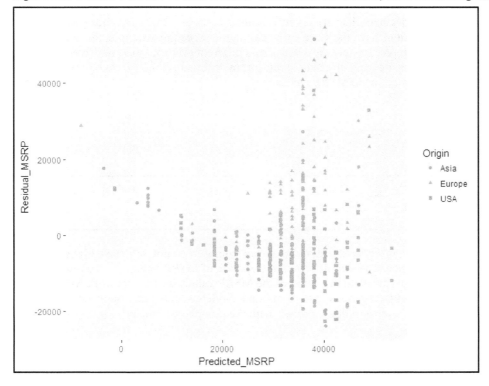

Let's continue to improve the linear regression model for prediction MSRP by adding more predictors to the model. In this example, we add three categorical predictors, Origin, Type, and DriveTrain, and two more continuous predictors, Weight and Length. The categorical predictors must be specified in both the inputs and the nominals parameters. The R-square statistic is improved again (0.7).

```
> nomList <- c('Origin','Type','DriveTrain')
> contList <- c('MPG_City','Weight','Length')
> cas.regression.glm(
+    cars, target = 'MSRP', inputs = c(nomList,contList), nominals = nomList,
+    display = list(names = 'FitStatistics'),
+    output = list(casOut = list(name='MSRPPrediction2', replace = TRUE),
+                  copyvars = 'all',
+                  pred  = 'Predicted_MSRP',
+                  resid = 'Residual_MSRP',
+                  lcl = 'LCL_MSRP',
+                  ucl = 'UCL_MSRP'
+    ))
```

```
$FitStatistics
     RowId Description      Value
1     RMSE    Root MSE 8.902096e+03
2  RSQUARE    R-Square 7.025591e-01
3   ADJRSQ    Adj R-Sq 6.945595e-01
4      AIC         AIC 8.092009e+03
5     AICC        AICC 8.092903e+03
6      SBC         SBC 7.717521e+03
7 TRAIN_ASE        ASE 7.698848e+07
```

A linear regression model with categorical effects still assumes homogeneity of variance (that is, the random errors for all observations have the same variance). In the preceding example, this means that the variation of MSRP values is approximately the same across different origins, types, or drivetrains. Sometimes data can be heterogeneous. For example, cars from different origins might have different MSRP values, as well as different variations in MSRP values.

```
> cars@where <- ''
> cars@groupby <- 'Origin'
> out <- cas.simple.summary(cars,inputs=c('MSRP'))
> > result <- rbind.bygroups(out)
> result$Summary[,c('Origin','Column','Mean','Var','Std')]
```

```
  Origin Column     Mean       Var      Std
1   Asia   MSRP 24741.32 128166619 11321.07
2 Europe   MSRP 48349.80 641031529 25318.60
3    USA   MSRP 28377.44 137170534 11711.98
```

The output from the simple.summary action shows that cars from Europe not only have higher MSRP values but also a greater variance. The sample standard deviation of MSRP values for the European cars doubles compared to the sample standard deviation of the MSRP values for the cars from Asia and USA. One easy remedy for variance heterogeneity is to fit multiple models, one for each segment of the data. For the regression.glm action, you can fit multiple models with the groupby parameter:

```
> cars@groupby <- 'Origin'
> cars@where <- 'MSRP < 100000 and MPG_City < 40'
> nomList <- c('Type','DriveTrain')
> contList <- c('MPG_City','Weight','Length')
```

```
> tmp <- cas.regression.glm(
+    cars, target = 'MSRP', inputs = c(nomList,contList), nominals = nomList,
+    display = list(names = "FitStatistics"),
+    output = list(casOut = list(name='MSRPPredictionGroupBy', replace = TRUE),
+                  copyvars = 'all',
+                  pred  = 'Predicted_MSRP',
+                  resid = 'Residual_MSRP',
+                  lcl = 'LCL_MSRP',
+                  ucl = 'UCL_MSRP'
+    ))
> groupBYResult <- defCasTable(conn, 'MSRPPredictionGroupBy')
> out <- cas.table.fetch(groupBYResult, to = 1000)$Fetch
> ggplot(out, aes(Predicted_MSRP, Residual_MSRP,
+                  colour = Origin, shape = Origin)) +
+    geom_point()
```

Figure 8.4 shows a scatter plot of predicted MSRP values and residuals with three linear regression models fit for each origin (Asia, Europe, and USA):

Figure 8.4: Scatter Plot of Predicted MSRP Values and Residuals with Three Linear Regression Models Fit for Each Origin

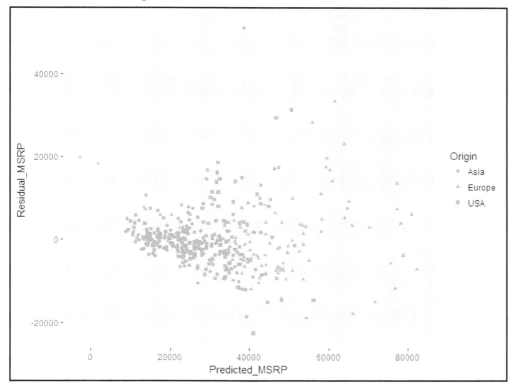

Extensions of Ordinary Linear Regression

The key assumptions of an ordinary linear regression are 1) the expected value of the dependent variable can be predicted using a linear combination of a set of predictors, and 2) the error distribution function of ε follows a normal distribution. The second assumption might not be valid for some applications, such as estimating the number of calls received within a time interval in a call center. Generalized linear models (GLM) and regression trees are popular types of generalization of ordinary linear regression to fit data that does not follow a normal distribution.

Generalized Linear Models

In generalized linear models, the dependent variable Y is assumed to follow a particular probability distribution, and the expected value of Y is predicted by a function of the linear combination of the predictors,

$$E(y) = f(x_1, x_2, \cdots, x_K) = f(a + b_1 x_1 + b_2 x_2 + \cdots + b_K x_K)$$

where $E(y)$ is the expected values of $Y, x_1, x_2, \cdots, x_K$ are the observed values of the predictors $X_1, X_2, \cdots, X_K$, and $b_1, b_2, \cdots, b_K$ are the unknown parameters. It is more common to express a GLM model by linking the expected values of Y to the linear combination of predictors,

$$g[E(y)] = a + b_1 x_1 + b_2 x_2 + \cdots + b_K x_K$$

The link function $g()$ is the inverse function of $f()$. The probability distribution of Y is usually from the exponential family of distribution, such as normal, binomial, exponential, gamma, Poisson, and zero-inflated distributions. The choice of the link function usually depends on the assumption of the probability distribution. For example, for call center data, it is common to assume that the number of calls within a time interval follows a Poisson distribution,

$$P(k\ calls\ in\ interval) = \frac{\lambda^k e^{-\lambda}}{k!}, k = 0, 1, 2 \ldots$$

and a log link function between the expected number of calls λ to the linear combination of the predictors.

$$\log(\lambda) = a + b_1 x_1 + b_2 x_2 + \cdots + b_K x_K$$

It is also worthwhile to mention that ordinary linear regression is a special type of GLM, where the target variable follows a normal distribution, and the link function is an identity function $E(y) = a + b_1 x_1 + b_2 x_2 + \cdots + b_K x_K$. For more details of generalized linear models, refer to [1] and [2].

Generalized linear models are available in the regression CAS action set. Let's continue to use the Cars data set example and build a simple generalized linear model to predict MSRP values of cars using MPG_City.

```
> cars@groupby <- ''
> cars@where <- ''
> cas.regression.genmod(
+    cars,
+    model = list(depVars = list(name = 'MSRP'),
+                 effects = c('MPG_City'),
+                 dist    = 'gamma',
+                 link    = 'log'
+                 )
+    )
```

```
NOTE: Convergence criterion (GCONV=1E-8) satisfied.
$ConvergenceStatus
                                        Reason Status
1 Convergence criterion (GCONV=1E-8) satisfied.      0
   MaxGradient
1 1.068416e-09


$Dimensions
          RowId               Description Value
1 NDESIGNCOLS         Columns in Design     2
2    NEFFECTS         Number of Effects     2
3   MAXEFCOLS         Max Effect Columns    1
4  DESIGNRANK           Rank of Design      2
5     OPTPARM Parameters in Optimization    3


$FitStatistics
  RowId            Description    Value
1  M2LL       -2 Log Likelihood 9270.853
2   AIC  AIC (smaller is better) 9276.853
3  AICC AICC (smaller is better) 9276.910
4   SBC  SBC (smaller is better) 9289.031


$ModelInfo
        RowId               Description
1        DATA               Data Source
2 RESPONSEVAR         Response Variable
3        DIST               Distribution
4        LINK             Link Function
5         TECH Optimization Technique
                              Value
1                              CARS
2                              MSRP
3                             Gamma
4                               Log
5 Newton-Raphson with Ridging


$NObs
  RowId                Description Value
1 NREAD Number of Observations Read   428
2 NUSED Number of Observations Used   428
```

```
$ParameterEstimates
        Effect   Parameter  ParmName DF   Estimate       StdErr
1   Intercept   Intercept  Intercept  1 11.3077899 0.059610593
2    MPG_City    MPG_City   MPG_City  1 -0.0473999 0.002800635
3 Dispersion  Dispersion Dispersion  1  5.8865740 0.391526276
        ChiSq    ProbChiSq
1 35983.9291 0.000000e+00
2   286.4454 2.958655e-64
3        NaN          NaN


$Timing
           RowId              Task         Time      RelTime
1          SETUP Setup and Parsing 3.292203e-02 3.364603e-01
2    LEVELIZATION      Levelization 2.941799e-02 3.006494e-01
3 INITIALIZATION Model Initialization 1.831007e-02 1.871273e-01
4           SSCP  SSCP Computation 6.412983e-03 6.554014e-02
5        FITTING    Model Fitting 1.020193e-02 1.042629e-01
6        CLEANUP          Cleanup 2.145767e-06 2.192956e-05
7          TOTAL            Total 9.784818e-02 1.000000e+00
```

In the preceding example, we fit a generalized linear model using gamma distribution and the log link function because MSRP values of cars are nonnegative continuous values. The type of distribution is usually determined by the range and the shape of the data. For example, exponential, gamma, and inverse Gaussian distributions are popular choices for fitting nonnegative continuous values, such as MSRP values of cars, sales revenue, insurance claim amounts, and so on. Binomial and multinomial distributions are a valid distribution assumption to fit an integer response. In the next example, we fix a multinomial regression model to predict the number of cylinders of the vehicles that are based on MPG_City.

```
> cas.regression.genmod(
+   cars,
+   model = list(depVars = list(name = 'Cylinders'),
+                effects = c('MPG_City'),
+                dist    = 'multinomial',
+                link    = 'logit'),
+   display = list(list(names = list('ModelInfo', 'ParameterEstimates'))))
+ )
```

```
NOTE: Convergence criterion (GCONV=1E-8) satisfied.

$ModelInfo
        RowId              Description                Value
1        DATA              Data Source                 CARS
2 RESPONSEVAR          Response Variable           Cylinders
3     NLEVELS Number of Response Levels                    7
4        DIST              Distribution          Multinomial
5    LINKTYPE                Link Type           Cumulative
6        LINK            Link Function                Logit
7        TECH   Optimization Technique Newton-Raphson with Ridging

$ParameterEstimates
      Effect Parameter    ParmName Outcome Cylinders DF    Estimate
1 Intercept Intercept Intercept_3       3         3  1  -60.329075
2 Intercept Intercept Intercept_4       4         4  1  -21.461149
3 Intercept Intercept Intercept_5       5         5  1  -21.233691
4 Intercept Intercept Intercept_6       6         6  1  -16.632445
5 Intercept Intercept Intercept_8       8         8  1  -10.988487
```

```
6 Intercept Intercept Intercept_10     10        10  1 -10.314220
7  MPG_City  MPG_City     MPG_City               NaN  1   1.013934

      StdErr     ChiSq    ProbChiSq
1 4.82953335 156.04253 8.286542e-36
2 1.58488726 183.36194 8.941488e-42
3 1.57576625 181.57975 2.190306e-41
4 1.33727470 154.69310 1.634032e-35
5 1.13947020  92.99719 5.236863e-22
6 1.18654121  75.56264 3.539969e-18
7 0.07737144 171.73470 3.092446e-39
```

In the preceding example, the cumulative logit link function is used for the multinomial model. For a generalized linear model using multinomial distribution, we estimate where an observation (a car) can fall into one of the seven possible number of cylinders (Cylinder=3, 4, 5, 6, 8, 10, 12). In this case, the cumulative log link assumes a logit link function between the cumulative probabilities and the linear combination of the predictors:

$$\Pr\left(\text{Cylinders} = 3\right) = f\left(-60.329075 + 1.013934 \times \text{MPG\_City}\right)$$

$$\Pr\left(\text{Cylinders} = 3\right) + \Pr\left(\text{Cylinders} = 4\right) = f\left(-21.461149 + 1.013934 \times \text{MPG\_City}\right)$$

$$......$$

$$\Pr\left(\text{Cylinders} = 3\right) + \cdots + \Pr\left(\text{Cylinders} = 10\right) = f\left(-10.314220 + 1.013934 \times \text{MPG\_City}\right)$$

where $f(u) = \dfrac{\exp(u)}{1 + \exp(u)}$ is the standard inverse logit link function. You can use the parameter estimates to score a new observation directly. For example, when a car has MPG_CITY = 20, the chance that this car is 4-cylinder is about 23.5%:

$$\Pr\left(\text{Cylinders} = 4 \mid \text{MPG\_CITY} = 20\right) = f\left(-21.461149 + 1.013934 \times 20\right) - f\left(-60.329075 + 1.013934 \times 20\right) = 0.235$$

Similar to the regression.glm action, the regression.genmod action generates the predictions from the generalized linear model. Instead of using the parameter estimates from the preceding formulas to manually compute the predictions, you can use the regression.genmod action to score a data set using the output options.

```
> out <- cas.regression.genmod(
+    cars,
+    model = list(depVars = list(name = 'Cylinders'),
+                 effects = c('MPG_City'),
+                 dist    = 'multinomial',
+                 link    = 'logit'),
+    output= list(casout  = list(name = 'CylinderPredicted',
+                                 replace = TRUE),
+                 copyVars= 'ALL',
+                 pred = 'Prob_Cylinders')
+
+ )
```

```
NOTE: Convergence criterion (GCONV=1E-8) satisfied.
```

```
> result <- defCasTable(conn, 'CylinderPredicted')
> head(result[[c('Prob_Cylinders','_LEVEL_','Cylinders','MPG_City')]], n = 24)
```

	Prob_Cylinders	_LEVEL_	Cylinders	MPG_City
1	1.928842e-19	3	6	17
2	1.442488e-02	4	6	17
3	1.804258e-02	5	6	17
4	6.466697e-01	6	6	17
5	9.980702e-01	8	6	17
6	9.990158e-01	10	6	17
7	2.331945e-16	3	4	24
8	9.465090e-01	4	4	24
9	9.569226e-01	5	4	24
10	9.995483e-01	6	4	24
11	9.999984e-01	8	4	24
12	9.999992e-01	10	4	24
13	3.069208e-17	3	4	22
14	6.996011e-01	4	4	22
15	7.451397e-01	5	4	22
16	9.965780e-01	6	4	22
17	9.999878e-01	8	4	22
18	9.999938e-01	10	4	22
19	4.039564e-18	3	6	20
20	2.346085e-01	4	6	20
21	2.778780e-01	5	6	20
22	9.745742e-01	6	6	20
23	9.999077e-01	8	6	20
24	9.999530e-01	10	6	20

You can generate the predictions for a car with MPG_CITY = 20 from the last six rows of the preceding output table:

$$\Pr(\text{Cylinders} = 3 \mid \text{MPG\_CITY} = 20) = 4.039564e - 18 \approx 0$$
$$\Pr(\text{Cylinders} = 4 \mid \text{MPG\_CITY} = 20) = 0.2346085 - 0 = 0.2346085$$
$$\Pr(\text{Cylinders} = 5 \mid \text{MPG\_CITY} = 20) = 0.2778780 - 0.2346085 = 0.04327$$
$$\Pr(\text{Cylinders} = 6 \mid \text{MPG\_CITY} = 20) = 0.9745742 - 0.2778780 = 0.696696$$
$$\Pr(\text{Cylinders} > 6 \mid \text{MPG\_CITY} = 20) = 1 - 0.9745742 < 0.03$$

Using the multinomial model, a car with MPG_CITY = 20 is most likely a 6-cylinder (69.7%) or 4-cylinder car (23.5%), and the chance that it has more than 6 cylinders is less than 3%.

Table 8.3 lists the output result tables from the regression.genmod action. You can use the display option to include or exclude a result table when fitting a generalized linear model using the regression.genmod action.

Table 8.3: Summary of the Output Result Tables from the regression.genmod Action

Table Name	Description
ModelInfo	Basic information about the model, such as the data source, the response variable, the distribution and link functions, and the optimization technique.
NObs	The number of observations that are read and used. Missing values are excluded, by default.
ConvergenceStatus	The convergence status of the parameter estimation.
Dimension	The dimension of the model, including the number of effects and the number of parameters.
FitStatistics	The fit statistics of the model such as log likelihood (multiplied by -2), AIC, AICC, and SBC.
ParameterEstimates	The estimation of the model parameters.
Timing	A timing of the subtasks of the GLM action call.

Regression Trees

Similar to decision tree models, regression trees are machine learning algorithms that use recursive partitioning to segment the input data set and to make predictions within each segment of the data. In this section, we briefly introduce how to use the tree models that are available in the decisiontree action set to build predictive models for continuous response. For information about decision tree and tree family models, see Chapter 9.

A regression tree usually follows these steps:

1. Grows a tree as deep as possible based on a splitting criterion and the training data.
2. Prunes back some nodes on the tree based on an error function on the validation data.
3. For each leaf (terminal node), build a simple model to predict the continuous dependent variable. The most common approach is to use the local sample average of the dependent variable.

Let's first load the decisiontree action set:

```
> cas.builtins.loadActionSet(conn, 'decisiontree')
```

```
NOTE: Added action set 'decisiontree'.
NOTE: Information for action set 'decisiontree':
NOTE:    decisionTree
NOTE:        dtreeTrain - Trains a decision tree
NOTE:        dtreeScore - Scores a table using a decision tree model
NOTE:        dtreeSplit - Splits decision tree nodes
NOTE:        dtreePrune - Prune a decision tree
NOTE:        dtreeMerge - Merges decision tree nodes
NOTE:        dtreeCode - Generates DATA step scoring code from a decision
             tree model
NOTE:        forestTrain - Trains a forest
NOTE:        forestScore - Scores a table using a forest model
NOTE:        forestCode - Generates DATA step scoring code from a forest model
NOTE:        gbtreeTrain - Trains a gradient boosting tree
NOTE:        gbtreeScore - Scores a table using a gradient boosting tree model
NOTE:        gbtreeCode - Generates DATA step scoring code from a gradient
             boosting tree model
```

The decisiontree.dtreetrain action fits the decision tree model for a categorical dependent variable or the regression tree model for a continuous dependent variable. Let's build a regression tree model using MPG_City to predict MSRP values of cars. In this example, we create the simplest regression tree that splits the root node into only two leaves (splitting the entire data into two partitions).

```
> result <- cas.decisiontree.dtreeTrain(
+    cars,
+    target = 'MSRP',
+    inputs = c('MPG_City'),
+    maxlevel = 2,
+    casout = list(name = 'treeModel1', replace = TRUE)
+    )
> output1 <- defCasTable(conn, 'treeModel1')
> output1[,c('_NodeID_', '_Parent_','_Mean_','_NodeName_',
+            '_PBLower0_','_PBUpper0_')]
```

	NodeID	_Parent_	_Mean_	_NodeName_	_PBLower0_
1	0	-1	32774.86	MPG_City	NaN
2	1	0	22875.34	MSRP	20
3	2	0	41623.09	MSRP	10

	PBUpper0
1	NaN
2	60
3	20

In the preceding example, the decision tree model is saved to the CAS table treeModel1. This table is stored on the CAS server. You can use the fetch or head method to download the table to the R client. Or if the CAS table is very small, you can define a CASTable object to reference it, and use column indexing as if you are using an R data.frame. The tree model table contains three observations, one for each node in the tree, including the root node. We fetch information only from the tree model table, such as the unique IDs of the nodes and parents (_NodeID_ and _Parent_), the local sample means of the dependent variable (_Mean_), the splitting variable (_NodeName_), and the splitting points (_PBLower0_ and _PBUpper0_). The tree model table treeModel1 also contains other useful information, such as the sizes of the nodes, the splitting criteria, and so on.

We can read about how the root node is split from the preceding table. The value of the_NodeName_ column in the first row shows that the root node is split by MPG_City. The values of the _PBLower0_ and _PBUpper0_ columns show us that Node 1 contains the observations with MPG_CITY in (20,60], and Node 2 contains the observations with MPG_CITY in [10,20]. Note that in the CAS tree models, the splitting points, such as MPG_CITY = 20 in this case are assigned to the child node with smaller values of MPG_City.

Based on this information, a tree structure follows for the first decision tree example shown in Figure 8.5.

Figure 8.5: Decision Tree Example

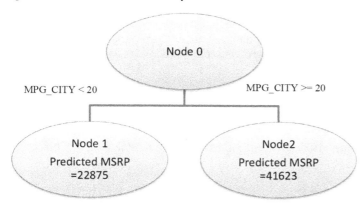

The splitting criteria of the CAS decision tree model are listed in Table 8.4.

Table 8.4: Splitting Criteria

Parameter	Description
crit = 'ftest'	Uses the p-values of F test to select the best split.
crit = 'variance'	Uses the best split that produces the largest reduction of sum of squared errors. For regression trees, the sum of squared errors is equivalent to the variable of the data.
crit = 'chaid'	Uses adjusted significance testing (Bonferroni testing) to select the best split.

Pruning a tree model is a necessary step to avoid overfitting when you use the regression tree model to score new data (validation data). In the preceding example, we grow only the simplest tree structure, so there is no need to prune it back. If you have created a deeper tree and you have validation data, you can use the decisiontree.dtreePrune action to prune a given tree model.

```
cas.decisiontree.dtreePrune(
  conn,
  table = 'your_vadliatoin_data',
  modelTable = 'treeModel1',
  casout = list(name = 'pruned_tree')
  )
```

The last step for completing a regression tree model is to use the pruned tree model to score your data. This is done by the decisiontree.dtreeScore action.

```
cas.decisiontree.dtreeScore(
  conn,
  table = 'your_vadliatoin_data',
  modelTable = 'pruned_tree'
  )
```

Conclusion

In this chapter, we introduced the linear regression model and discussed some best practices to improve the model fitting of a linear regression. Then we introduced the generalized linear model that is available in the regression action set and the actions in the decisiontree action set to build a regression tree model.

References

[1] Neter, John, et al. 1996. *Applied Linear Statistical Models*. Vol. 4. Chicago: Irwin.

[2] McCullagh, Peter, and John A. Nelder. 1989. *Generalized Linear Models*. Vol. 37. CRC Press.

Chapter 9: Modeling Categorical Variables

Overview

In Chapter 8, we introduced linear regression, generalized linear models, and regression trees for modeling continuous response variables. In this chapter, we focus on applications in which the response variable is categorical, such as organic food that is purchased in a supermarket (Bought, Not), blood pressure status (High, Normal, Low), and credit card application status (Accepted, Rejected). Logistic regression and decision trees are introduced in the first two sections for a binary response variable, which is a response variable with only two qualitative outcomes. In the last section, we introduce random forests, gradient boosting trees, and neural networks, which can fit a categorical response variable with more than two distinct outcomes.

We continue to use the Organics data set for this chapter. Again, we assume that a CAS server is already set up and that the data sets have been loaded to the CAS server.

```
> cas.table.tableInfo(conn)
$TableInfo
               Name    Rows Columns IndexedColumns Encoding
1          ORGANICS 1688948      36              0    utf-8
  CreateTimeFormatted    ModTimeFormatted AccessTimeFormatted
1 10Dec2017:16:33:21 10Dec2017:16:33:21  10Dec2017:16:33:21
  JavaCharSet CreateTime     ModTime AccessTime Global Repeated View
1        UTF8 1828542801 1828542801 1828542801      0        0    0
        ourceName SourceCaslib Compressed Creator Modifier
1 organics.sashdat    yourcaslib          0 username
```

Logistic Regression

Similar to the linear regression models introduced in Chapter 8, logistic regression uses the linear combination of one or more predictors to build a predictive model for the response outcomes. Unlike linear regression, the prediction of interest in a logistic regression is not a continuous outcome with a range from negative to positive infinity. Instead, we are interested in predicting the probabilities of the outcome levels of the response variable. Examples are the probability of buying or not buying organic food from a supermarket or the probability of a credit card application that is accepted or rejected. For a binary response variable, the two distinct outcome levels are usually called the event level and the non-event level.

$$P_{event} + P_{non-event} = 1$$

We simply need to predict P_{event}, the event probability for the response variable, which always takes values in $[0, 1]$. Logistic regression originates from the logistic transformation, which maps a continuous input from $(-\infty, \infty)$ to an output between 0 and 1.

$$f(x) = \frac{1}{1 + e^{-k(x - x_0)}}$$

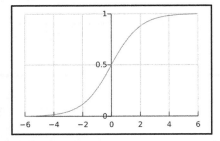

Similarly, a logistic regression model uses the logistic transformation to link the linear combination of one or more predictors to the probability of the event level as follows:

$$P_{event} = \frac{1}{1 + e^{-t}}$$

where t is a linear combination of the predictors:

$$t = a + b_1 x_1 + b_2 x_2 + \cdots + b_K x_K$$

The inverse form of the transformation also indicates that a logistic regression is a non-linear model:

$$G(P_{event}) = log\left(\frac{P_{event}}{1 - P_{event}}\right) = a + b_1 x_1 + b_2 x_2 + \cdots + b_K x_K$$

where the function $G(P_{event})$ is often called the logit link function. The ratio $P_{event}/1 - P_{event}$ is also known as the odds ratio. When a logit link function is used, we assume that the predictors have a linear relationship with the log of the odds ratio. Logistic regression can be extended by using other link functions as well, such as probit, cloglog, and negative cloglog.

Logistic regression models are available in the regression action set. First, you must load the action set.

```
> cas.builtitns.loadActionSet(conn, 'regression')
```

```
NOTE: Added action set 'regression'.
NOTE: Information for action set 'regression':
NOTE:    regression
NOTE:        glm - Fits linear regression models using the method of least
             squares
NOTE:        genmod - Fits generalized linear regression models
NOTE:        logistic - Fits logistic regression models
NOTE:        genmodScore - creates a table on the server that contains
             results from scoring observations by using a fitted model.
NOTE:        logisticScore - creates a table on the server that contains
             results from scoring observations by using a fitted model.
NOTE:        glmScore - creates a table on the server that contains results
             from scoring observations by using a fitted model.
```

Now let's build a simple logistic regression from the Organics data set to predict whether consumers like to buy organic food. The response variable in this data set is called TargetBuy.

```
> cas.regression.logistic(
+    organics,
+    target = 'TargetBuy',
+    inputs = c('DemAge', 'Purchase_3mon', 'Purchase_6mon')
+ )
```

```
NOTE: Added action set 'regression'.
NOTE: Convergence criterion (GCONV=1E-8) satisfied.

$ConvergenceStatus
                                          Reason Status  MaxGradient
1 Convergence criterion (GCONV=1E-8) satisfied.      0 4.419883e-08

$Dimensions
        RowId                Description Value
1 NDESIGNCOLS         Columns in Design     4
2    NEFFECTS         Number of Effects     4
3   MAXEFCOLS       Max Effect Columns      1
4  DESIGNRANK           Rank of Design     4
5     OPTPARM Parameters in Optimization    4

$FitStatistics
  RowId           Description   Value
1  M2LL       -2 Log Likelihood 1608090
2   AIC  AIC (smaller is better) 1608098
3  AICC AICC (smaller is better) 1608098
4   SBC  SBC (smaller is better) 1608147

$GlobalTest
              Test DF   ChiSq ProbChiSq
1 Likelihood Ratio  3 149251.1        0

$ModelInfo
        RowId           Description                Value
1         DATA           Data Source              ORGANICS
2 RESPONSEVAR      Response Variable             TargetBuy
3         DIST          Distribution                Binary
```

```
4              LINK          Link Function                        Logit
5              TECH Optimization Technique Newton-Raphson with Ridging

$NObs
  RowId                    Description    Value
1 NREAD Number of Observations Read 1688948
2 NUSED Number of Observations Used 1574340

$ParameterEstimates
           Effect      Parameter      ParmName DF      Estimate
1      Intercept      Intercept      Intercept  1  1.755274e+00
2         DemAge         DemAge         DemAge  1 -5.743798e-02
3 purchase_3mon purchase_3mon purchase_3mon  1 -1.617231e-06
4 purchase_6mon purchase_6mon purchase_6mon  1  3.872414e-05
        StdErr         ChiSq       ProbChiSq
1 5.709205e-02 9.452326e+02 1.442299e-207
2 1.584786e-04 1.313582e+05  0.000000e+00
3 5.495954e-05 8.658809e-04  9.765250e-01
4 3.890189e-05 9.908827e-01  3.195267e-01

$ResponseProfile
  OrderedValue Outcome TargetBuy    Freq Modeled
1            1  Bought    Bought  387600       *
2            2      No        No 1186740

$Timing
          RowId                Task         Time       RelTime
1          SETUP   Setup and Parsing 1.032121e-01 2.938332e-01
2   LEVELIZATION        Levelization 9.885693e-02 2.814345e-01
3 INITIALIZATION Model Initialization 4.235697e-02 1.205855e-01
4           SSCP    SSCP Computation 6.504059e-03 1.851632e-02
5        FITTING       Model Fitting 9.797502e-02 2.789238e-01
6        CLEANUP             Cleanup 2.861023e-06 8.145008e-06
7          TOTAL               Total 3.512609e-01 1.000000e+00
```

In the preceding example, we build a logistic regression to predict the probability of buying organic food (Pr(TargetBuy = Bought)) using three continuous predictors, Age, Recent 3-Month Purchase Amount, and Recent 6-month Purchase Amount. You can also add categorical predictors to the logistic regression. They must be specified in both the inputs option and in the nominals option.

```
> cas.regression.logistic(
+   organics,
+   target = 'TargetBuy',
+   inputs = c('DemAge', 'Purchase_3mon', 'Purchase_6mon',
+              'DemGender','DemHomeowner'),
+   nominals = c('DemGender', 'DemHomeowner'),
+   display = list(list(names = 'ParameterEstimates'))
+ )
```

```
NOTE: Convergence criterion (GCONV=1E-8) satisfied.
$ParameterEstimates
          Effect DemGender DemHomeowner          Parameter
1      Intercept                                  Intercept
2         DemAge                                     DemAge
3 purchase_3mon                              purchase_3mon
4 purchase_6mon                              purchase_6mon
5      DemGender         F                          DemGender F
```

```
6        DemGender          M                      DemGender M
7        DemGender          U                      DemGender U
8     DemHomeowner                     No    DemHomeowner No
9     DemHomeowner                    Yes DemHomeowner Yes
          ParmName DF     Estimate        StdErr        ChiSq
1        Intercept  1  3.533485e-01 5.958582e-02 3.516582e+01
2           DemAge  1 -5.647790e-02 1.624806e-04 1.208242e+05
3     purchase_3mon  1  8.864158e-06 5.694667e-05 2.422916e-02
4     purchase_6mon  1  3.650636e-05 4.033218e-05 8.192828e-01
5      DemGender_F  1  1.817158e+00 7.381156e-03 6.060891e+04
6      DemGender_M  1  8.579051e-01 8.215534e-03 1.090453e+04
7      DemGender_U  0  0.000000e+00          NaN          NaN
8  DemHomeowner_No  1  3.197861e-04 4.226444e-03 5.724915e-03
9 DemHomeowner_Yes  0  0.000000e+00          NaN          NaN
      ProbChiSq
1 3.027918e-09
2 0.000000e+00
3 8.763032e-01
4 3.653900e-01
5 0.000000e+00
6 0.000000e+00
7          NaN
8 9.396871e-01
9          NaN
```

In the preceding example, we also use the display option to select only the parameter estimate output table. For each categorical predictor, the logistic action constructs dummy indicators for the distinct levels of the predictor, and uses them as linear terms. For example, if a consumer is 30 years old, has purchased the amount $1,200 and $500 for the last six months and three months, respectively, is male, and currently does not own a home, the predicted probability of buying organic food is 39.65%.

$$t = 0.3533 - 0.056 \times 30 + 0.000008864 \times 500 + 0.00003651 \times 1200 + 0.8579 + 0.0003197 = -0.42024$$

$$P_{event} = \frac{1}{1 + e^{-t}} = 39.65\%$$

Table 9.1 summarizes the output tables that are returned by the regression.logistic action.

Table 9.1: Summary of Output Tables Returned by the regression.logistic Action

Table Name	Description
NObs	The number of observations that are read and used. Missing values are excluded, by default.
ResponseProfile	The frequency distribution of the response variable. The event level is marked with an asterisk (*).
ConvergenceStatus	The convergence status of the parameter estimation.
Dimension	The dimension of the model, including the number of effects and the number of parameters.
GlobalTest	A likelihood ratio test that measures the overall model fitting.
FitStatistics	The fit statistics of the model, such as log likelihood (multiplied by 2), AIC, AICC, and BIC.
ParameterEstimates	The estimation of the logistic regression parameters.
Timing	A timing of the subtasks of the GLM action call.

In the preceding example, first, we define a simple logistic model including the class effect list (nominals), the response variables (target), and the effect list (inputs). If you want to define a more complicated logistic regression model, you need to use another set of parameters. You need to put the model-related parameters inside the model option. The regression.logistic action, by default, uses the logit link function, but you can always override it with your own selection. In the following example, we change the link function to PROBIT:

```
> cas.regression.logistic(
+   organics,
+   model = list(
+     depvars = list(list(name = 'TargetBuy')),
+     effects = list(list(
+       vars = list('DemAge', 'Purchase_3mon', 'Purchase_6mon',
+                   'DemGender','DemHomeowner'))),
+     link = 'PROBIT'),
+   class = list(vars = list('DemGender', 'DemHomeowner')),
+   display = list(list(names = list('ResponseProfile',
+                                    'ParameterEstimates')))
+ )
```

```
NOTE: Convergence criterion (GCONV=1E-8) satisfied.
$ParameterEstimates
          Effect DemGender DemHomeowner              Parameter
1      Intercept                                     Intercept
2         DemAge                                        DemAge
3 purchase_3mon                                 purchase_3mon
4 purchase_6mon                                 purchase_6mon
5      DemGender         F                         DemGender F
6      DemGender         M                         DemGender M
7      DemGender         U                         DemGender U
8  DemHomeowner                   No    DemHomeowner No
9  DemHomeowner                   Yes DemHomeowner Yes
          ParmName DF      Estimate         StdErr          ChiSq
1        Intercept  1  1.715821e-01 3.445468e-02 2.479977e+01
2           DemAge  1 -3.153958e-02 8.963853e-05 1.238005e+05
3    purchase_3mon  1  6.638325e-06 3.299864e-05 4.046922e-02
4    purchase_6mon  1  2.108883e-05 2.336458e-05 8.146839e-01
5      DemGender_F  1  1.011867e+00 3.790977e-03 7.124339e+04
6      DemGender_M  1  4.538139e-01 4.258421e-03 1.135686e+04
7      DemGender_U  0  0.000000e+00          NaN          NaN
8  DemHomeowner_No  1  5.764126e-04 2.448504e-03 5.541985e-02
9 DemHomeowner_Yes  0  0.000000e+00          NaN          NaN
     ProbChiSq
1 6.360516e-07
2 0.000000e+00
3 8.405659e-01
4 3.667391e-01
5 0.000000e+00
6 0.000000e+00
7          NaN
8 8.138873e-01
9          NaN

$ResponseProfile
  OrderedValue Outcome TargetBuy    Freq Modeled
1            1  Bought    Bought  387600       *
2            2      No        No 1186740
```

The logistic regression models that we have built so far train the model without actually using the model to make a prediction (that is, estimating $P(event)$). You must specify an output table for the logistic action in order to score the observations in the input table. By default, the casout table contains only the predictions, but you can use the copyvars option to copy some columns from the input table to the casout table. You can use copyVars='ALL' to copy all columns to the casout table.

```
> cas.regression.logistic(
+    organics,
+    model = list(
+      depvars = list(list(name = 'TargetBuy')),
+      effects = list(list(
+        vars = list('DemAge', 'Purchase_3mon', 'Purchase_6mon',
+                    'DemGender','DemHomeowner'))),
+      link = 'PROBIT'),
+    class = list(vars = list('DemGender', 'DemHomeowner')),
+    output = list(casout  = list(name = 'predicted',
+                                 replace = TRUE),
+                  copyVars= 'ALL')
+    )
> result1 = defCasTable(conn, 'predicted')
```

The regression.logistic action generates one more result table when a casout table is created. This table contains basic information about the casout table, such as the name, the library, and the dimensions of the table.

```
$OutputCasTables

                  casLib       Name Label     Rows     Columns
 1   CASUSERHDFS(username)   predicted        1688948      37
```

You can print out the column names of the output table to see what new columns have been added. By default, the regression.logistic action creates a new column _PRED_ for the estimated event probability.

```
> names(result1)
```

```
 [1]  "_PRED_"              "ID"                  "DemAffl"
 [4]  "DemAge"              "DemGender"           "DemHomeowner"
 [7]  "DemAgeGroup"         "DemCluster"          "DemReg"
[10]  "DemTVReg"            "DemFlag1"            "DemFlag2"
[13]  "DemFlag3"            "DemFlag4"            "DemFlag5"
[16]  "DemFlag6"            "DemFlag7"            "DemFlag8"
[19]  "PromClass"           "PromTime"            "TargetBuy"
[22]  "Bought_Beverages"    "Bought_Bakery"       "Bought_Canned"
[25]  "Bought_Dairy"        "Bought_Baking"       "Bought_Frozen"
[28]  "Bought_Meat"         "Bought_Fruits"       "Bought_Vegetables"
[31]  "Bought_Cleaners"     "Bought_PaperGoods"   "Bought_Others"
[34]  "purchase_3mon"       "purchase_6mon"       "purchase_9mon"
[37]  "purchase_12mon"
```

You can use the data summary actions in the simple action set to look at the model output. For example, let's compare the average predicted probability of buying organic foods across different levels of DemGender:

```
> cas.simple.crossTab(result1, row='DemGender', weight='_PRED_', aggregators='mean')
```

```
$Crosstab
  DemGender       Col1
1          F 0.34513810
2          M 0.16670282
3          U 0.07819483
```

From the preceding output, it looks like female customers are more likely to purchase organic food, and the customers who didn't provide gender information are not interested in buying organic food. The ratio (0.343891/0.076981) between these two groups is approximately 5, which means that female customers are five times more likely to purchase organic foods than the customers who didn't provide gender information.

The regression.logistic action generates additional model diagnosis outputs. Table 9.2 summarizes the columns that a regression.logistic action can generate.

Table 9.2: Summary of the Columns that regression.logistic Action Can Generate

Option	Description
pred	The predicted value. If you do not specify any output statistics, the predicted value is named PRED, by default.
resraw	The raw residual.
xbeta	The linear predictor.
stdxbeta	The standard error of the linear predictor.
lcl	The lower bound of a confidence interval for the linear predictor.
ucl	The upper bound of a confidence interval for the linear predictor.
lclm	The lower bound of a confidence interval for the mean.
uclm	The upper bound of a confidence interval for the mean.
h	The leverage of the observation.
reschi	The Pearson chi-square residual.
stdreschi	The standardized Pearson chi-square residual.
resdev	The deviance residual.
reslik	The likelihood residual (likelihood displacement).
reswork	The working residual.
difdev	The change in the deviance that is attributable to deleting the individual observation.
difchisq	The change in the Pearson chi-square statistic that is attributable to deleting the individual observation.
cbar	The confidence interval displacement, which measures the overall change in the global regression estimates due to deleting the individual observation.
alpha	The significance level used for the construction of confidence intervals.

Similar to the code option in the regression.glm action for linear regression, SAS DATA step code can also be used to save a logistic model. You can score new data sets using the DATA step code in R (using the runcode action from the datastep action set) or in a SAS language environment, such as SAS Display Manager, SAS Enterprise Guide, and SAS Studio.

```
> # example 4 score code
> result <- cas.regression.logistic(
+    organics,
+    target = 'TargetBuy',
```

```
+    inputs = c('DemAge', 'Purchase_3mon', 'Purchase_6mon'),
+    code = list(tabForm = FALSE)
+ )
```

```
NOTE: Convergence criterion (GCONV=1E-8) satisfied.
```

```
> result$`_code_`
```

```
                                                          SASCode
1       /*---------------------------------------------------------
2                                   Generated SAS Scoring Code
3                                      Date: 10Dec2017:23:23:59
4         ---------------------------------------------------*/
5
6                         drop _badval_ _linp_ _temp_ _i_ _j_;
7                                             _badval_ = 0;
8                                             _linp_   = 0;
9                                             _temp_   = 0;
10                                            _i_      = 0;
11                                            _j_      = 0;
12                                          drop MACLOGBIG;
13                      MACLOGBIG= 7.0978271289338392e+02;
14
15                          array _xrow_0_0_{4} _temporary_;
16      array _beta_0_0_{4} _temporary_ (    1.75527422456685
17                                          -0.05743797902413
18                                          -1.6172314911813E-6
19                                           0.00003872414134);
20
21                              if missing(purchase_3mon)
22                                      or missing(DemAge)
23                              or missing(purchase_6mon)
24                                                then do;
25                                           _badval_ = 1;
26                                           goto skip_0_0;
27                                                     end;
28
29              do _i_=1 to 4; _xrow_0_0_{_i_} = 0; end;
30
31                                      _xrow_0_0_[1] = 1;
32
33                                 _xrow_0_0_[2] = DemAge;
34
35                         _xrow_0_0_[3] = purchase_3mon;
36
37                         _xrow_0_0_[4] = purchase_6mon;
38
39                                          do _i_=1 to 4;
40      _linp_ + _xrow_0_0_{_i_} * _beta_0_0_{_i_};
41                                                     end;
42
43                                               skip_0_0:
44                              length I_TargetBuy $6;
45              label I_TargetBuy = 'Into: TargetBuy';
46              array _levels_0_{2} $ _TEMPORARY_ ('Bought'
47                                                      ,'No'
48                                                         );
49              label P_TargetBuy = 'Predicted: TargetBuy';
```

```
50              if (_badval_ eq 0) and not missing(_linp_) then do;
51                                  if (_linp_ > 0) then do;
52              P_TargetBuy = 1 / (1+exp(-_linp_));
53                                         end; else do;
54          P_TargetBuy = exp(_linp_) / (1+exp(_linp_));
55                                                  end;
56          if P_TargetBuy >= 0.5              then do;
57                          I_TargetBuy = _levels_0_{1};
58                                         end; else do;
59                          I_TargetBuy = _levels_0_{2};
60                                                  end;
61                                    end; else do;
62                                          _linp_ = .;
63                                     P_TargetBuy = .;
64                                   I_TargetBuy = ' ';
65                                                  end;
```

CAS also provides several useful data set options that can interact with an analytical action. For example, to build multiple logistic regressions (one for each level of a categorical variable), you can simply use the groupby option for the Organics data set.

```
> organics@groupby <- 'DemGender'
> result <- cas.regression.logistic(
+   organics,
+   target = 'TargetBuy',
+   inputs = c('DemAge', 'Purchase_3mon', 'Purchase_6mon')
+ )
```

```
NOTE: Convergence criterion (GCONV=1E-8) satisfied.
NOTE: Convergence criterion (GCONV=1E-8) satisfied.
NOTE: Convergence criterion (GCONV=1E-8) satisfied.
```

There are three convergence messages for the three logistic models that are trained for the female customers, male customers, and customers who didn't provide gender information. You can use the rbind.bygroups built-in function to concatenate the results and generate the parameter estimations:

```
> result2 <- rbind.bygroups(result)
> result2$ParameterEstimates
```

	DemGender	Effect	Parameter	ParmName	DF
1	F	Intercept	Intercept	Intercept	1
2	F	DemAge	DemAge	DemAge	1
3	F	purchase_3mon	purchase_3mon	purchase_3mon	1
4	F	purchase_6mon	purchase_6mon	purchase_6mon	1
5	M	Intercept	Intercept	Intercept	1
6	M	DemAge	DemAge	DemAge	1
7	M	purchase_3mon	purchase_3mon	purchase_3mon	1
8	M	purchase_6mon	purchase_6mon	purchase_6mon	1
9	U	Intercept	Intercept	Intercept	1
10	U	DemAge	DemAge	DemAge	1
11	U	purchase_3mon	purchase_3mon	purchase_3mon	1
12	U	purchase_6mon	purchase_6mon	purchase_6mon	1

	Estimate	StdErr	ChiSq	ProbChiSq
1	2.181055e+00	7.077208e-02	9.497499e+02	1.503554e-208
2	-5.701612e-02	1.922815e-04	8.792658e+04	0.000000e+00
3	2.329004e-05	6.805278e-05	1.171248e-01	7.321740e-01

```
4    3.788999e-05 4.824883e-05 6.167020e-01  4.322755e-01
5    1.230455e+00 1.272447e-01 9.350871e+01  4.044172e-22
6   -5.611447e-02 3.604855e-04 2.423116e+04  0.000000e+00
7   -9.149473e-05 1.227469e-04 5.556110e-01  4.560341e-01
8    6.508839e-05 8.668951e-05 5.637337e-01  4.527598e-01
9    2.113357e-01 2.043550e-01 1.069486e+00  3.010615e-01
10  -5.262068e-02 5.690589e-04 8.550638e+03  0.000000e+00
11   1.391191e-04 1.968059e-04 4.996857e-01  4.796382e-01
12  -4.563116e-05 1.391582e-04 1.075240e-01  7.429809e-01
```

Note that the groupby option in the R interface for SAS Viya is very similar to the BY statement in the SAS language for by-group processing of repeated analytical contents. However, the BY statement requires the data set to be sorted by the variables that are specified in the BY statement. The groupby option does not require pre-sorted data. When the data set is distributed, the groupby option does not introduce additional data shuffling either.

Decision Trees

A decision tree is a type of machine learning algorithm that uses recursive partitioning to segment the input data set and to make predictions within each segment of the data. Decision tree models have been widely used for predictive modeling, data stratification, missing value imputation, outlier detection and description, variable selection, and other areas of machine learning and statistical modeling. In this section, we focus on using decision trees to build predictive models.

Figure 9.1 illustrates a simple decision tree model on data that is collected in a city park regarding whether people purchase ice cream. The response variable is binary: 1 for the people who bought ice cream and 0 for the people who didn't buy ice cream. There are three predictors in this example: Sunny and Hot?, Have Extra Money?, and Crave Ice Cream?. In this example, all of these predictors are binary, with values YES and NO. In the example, the chance is 80% that people who have extra money on a sunny and hot day will buy ice cream.

Figure 9.1: Ice Cream Decision Tree Model

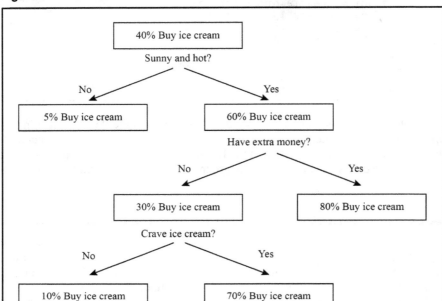

A typical decision tree model for a categorical variable contains these three steps:

1. Grow a decision tree as deep as possible based on a splitting criterion and the training data.
2. Prune back some nodes on the decision tree based on an error function on the validation data.
3. For each leaf, the probability of a response outcome is estimated using the sample frequency table.

The CAS decision tree supports a variety of decision tree splitting and pruning criteria. The decision tree action set contains machine learning models that are in the tree family: decision trees, random forests, and gradient boosting. Let's first load the decisiontree action set.

```
> cas.builtins.loadActionSet(conn, 'decisiontree')
```

```
NOTE: Added action set 'decisiontree'.
NOTE: Information for action set 'decisiontree':
NOTE:     decisionTree
NOTE:        dtreeTrain - Trains a decision tree
NOTE:        dtreeScore - Scores a table using a decision tree model
NOTE:        dtreeSplit - Splits decision tree nodes
NOTE:        dtreePrune - Prune a decision tree
NOTE:        dtreeMerge - Merges decision tree nodes
NOTE:        dtreeCode - Generates DATA step scoring code from a decision
                        tree model
NOTE:        forestTrain - Trains a forest
NOTE:        forestScore - Scores a table using a forest model
NOTE:        forestCode - Generates DATA step scoring code from a forest
                        model
NOTE:        gbtreeTrain - Trains a gradient boosting tree
```

```
NOTE:          gbtreeScore - Scores a table using a gradient boosting tree
                             model
NOTE:          gbtreeCode - Generates DATA step scoring code from a gradient
                             boosting tree model
```

The action sets with prefix dtree are used for building a decision tree model. The decisiontree.dtreeTrain action is used for training a decision tree. The decisiontree.dtreeScore and decisiontree.dtreeCode actions are designed for scoring using a decision tree and for generating as output a decision tree model, as SAS DATA step code. The other actions with the prefix dtree are used in interactive decision tree modifications, where a user can manually split, prune, or merge tree nodes.

Let's get started with a simple decision tree model that contains only one response variable and one predictor.

```
> cas.decisiontree.dtreeTrain(
+    organics,
+    target = 'TargetBuy',
+    inputs = c('DemGender'),
+    casout = list(name = 'treeModel1', replace = TRUE)
+ )
```

```
NOTE: Added action set 'decisiontree'.
$ModelInfo
                          Descr       Value
1            Number of Tree Nodes 5.000000e+00
2          Max Number of Branches 2.000000e+00
3                Number of Levels 3.000000e+00
4                Number of Leaves 3.000000e+00
5                  Number of Bins 2.000000e+01
6           Minimum Size of Leaves 3.236840e+05
7           Maximum Size of Leaves 9.233240e+05
8             Number of Variables 1.000000e+00
9   Confidence Level for Pruning 2.500000e-01
10   Number of Observations Used 1.688948e+06
11  Misclassification Error (%) 2.477163e+01

$OutputCasTables
                    casLib       Name Rows Columns
1   CASUSERHDFS(username)   treeModel1    5      24
```

The decisiontree.dtreeTrain action trains a decision tree model and saves it to the casout table treemodel1. The action also generates a model information table (ModelInfo) that contains basic descriptions of the trained decision tree model. Note that ModelInfo is stored at the local R client and contains only basic information about the tree model, whereas the treemodel1 table is a CAS table that is stored on the CAS server. Let's look at what is included in the tree model output:

```
> names(output1)
```

```
 [1] "_Target_"          "_NumTargetLevel_" "_TargetValL_"
 [4] "_TargetVal0_"      "_TargetVal1_"     "_CI0_"
 [7] "_CI1_"             "_NodeID_"         "_TreeLevel_"
[10] "_NodeName_"        "_Parent_"         "_ParentName_"
[13] "_NodeType_"        "_Gain_"           "_NumObs_"
[16] "_TargetValue_"     "_NumChild_"       "_ChildID0_"
```

```
[19] "_ChildID1_"         "_PBranches_"        "_PBNameL0_"
[22] "_PBNameL1_"         "_PBName0_"          "_PBName1_"
```

The decision tree output contains one row per node to summarize the tree structure and the model fit. Let's first fetch some columns that are related to the tree structure and the splitting values.

```
> output1[,c('_TreeLevel_', '_NodeID_', '_Parent_', '_ParentName_',
+            '_NodeType_', '_PBName0_',
+            '_PBName1_')]
```

```
  _TreeLevel_  _NodeID_  _Parent_  _ParentName_  _NodeType_  _PBName0_
1          0         0        -1                          1
2          1         1         0    DemGender            1          M
3          1         2         0    DemGender            3          F
4          2         3         1    DemGender            3          U
5          2         4         1    DemGender            3          M
  _PBName1_
1
2          U
3
4
5
```

The first column _TreeLevel_ identifies the depth of the tree nodes, where depth 0 indicates the root node. Each node has a unique node ID (_NodeID_) that is based on the order that the nodes are inserted in the decision tree. By default, the CAS decision tree model splits a node into two branches, which explains why the two nodes share the same parent (_Parent_).

The _ParentName_ column indicates which predictor is chosen to split its parent node. The columns with the prefix _PBName specify the splitting rule, which explains how to split the node from its parent node. For example, the second row and the third row tell the first split of the decision tree to split observations with DemGender = M/U and observations DemGender = F into two nodes with _NodeID_ 1 and 2. The _NodeType_ column in this table indicates whether a node is an internal node or a terminal node (leaf).

Figure 9.2 reconstructs the decision tree structure from the preceding fetched output.

Figure 9.2: Decision Tree Structure Reconstructed from the Fetched Output

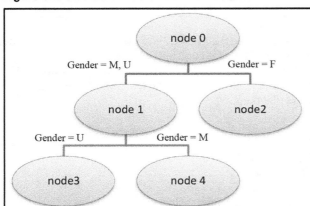

Next, let's look at the distribution of the target variables and the measurement of the splits.

```
> output1[,c('_TreeLevel_', '_NodeID_', '_Parent_',
+             '_TargetVal0_', '_TargetVal1_', '_CI0_', '_CI1_',
+             '_Gain_', '_NumObs_')]
```

TreeLevel	_NodeID_	_Parent_	_TargetVal0_	_TargetVal1_	_CI0_
1	0	0	-1	Bought	No 0.24771633
2	1	1	0	Bought	No 0.12904507
3	1	2	0	Bought	No 0.34611902
4	2	3	1	Bought	No 0.07842216
5	2	4	1	Bought	No 0.16612210

	_CI1_	_Gain_	_NumObs_
1	0.7522837	0.04771340	1688948
2	0.8709549	0.01288632	765624
3	0.6538810	0.00000000	923324
4	0.9215778	0.00000000	323684
5	0.8338779	0.00000000	441940

The first row returns the overall frequency distribution of the response variable. In this case, the overall sample percentage of those who buy organic food is 24.77%. The decision tree algorithm then evaluates all possible ways to cut the entire data into two segments. Because we have only one predictor with three levels, there are only three ways to split the data into two segments:

- Gender = F versus Gender = M, U
- Gender = M versus Gender = F, U
- Gender = U versus Gender = F, M

The decisiontree.dtreeTrain action evaluates all these three possible splits and selects the best one based on a certain criterion. The default criterion that is used by dtreetrain is information gain, which is the difference between the entropy of the parent node and the weighted average of the entropies of the child nodes. In other words, information gain tries to find a split that maximizes the difference of the target distribution between the parent node and the child nodes. The best split chosen by dtreetrain is Gender = F versus Gender = M, U with information gain 0.0477. The sample percentage of those who buy organic food and who are the two children of node 0 is 12.90% and 34.61%, respectively.

The tree model continues to evaluate the splits on the two segments and to determine whether they can be further split. The continued evaluation for more splits is the reason why a decision tree is also known as a recursive partition. The tree structure is grown using recursive partitioning, until one of the following criteria is met:

- The current partition cannot be split anymore.
- The minimum leaf size has been reached.
- The maximum depth of the tree has been reached.

Figure 9.3 shows the final decision tree model that we have built.

Figure 9.3: Final Decision Tree Model

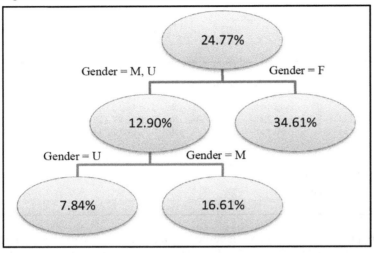

Table 9.3 lists some important parameters for training a decision tree model.

Table 9.3: Summary of Important Parameters for a Decision Tree

Parameter	Description	Default
maxbranch	Specifies the maximum number of children (branches) that are allowed for each level of the tree.	2
maxlevel	Specifies the maximum number of tree levels.	6
leafsize	Specifies the minimum number of observations on each node.	1
prune=TRUE | FALSE	Specifies whether to use a C4.5 pruning method for classification trees or minimal cost-complexity pruning for regression trees.	FALSE
varimp=TRUE | FALSE	Specifies whether the variable importance information is generated. The importance value is determined by the total Gini reduction.	FALSE
cflev	Specifies the aggressiveness of tree pruning according to the C4.5 method.	

The decisiontree.dtreeTrain action supports the following criteria shown in Table 9.4 for evaluating the splits in a decision tree growing process.

Table 9.4: Summary of Splitting Criteria

Parameter	Description
crit=chaid	CHAID (Chi-square Automatic Interaction Detector) technique
crit=chisquare	Chi-square test
crit=gain	Information gain
crit=gainratio	Information gain ratio
crit=gini	Per-leaf Gini statistic

Building a decision tree model usually involves a pruning step after the tree is grown. You must enable the prune option to request dtreetrain to include an additional pruning step after the tree is grown.

```
> cas.decisiontree.dtreeTrain(
+    organics,
+    target = 'TargetBuy',
+    inputs = c('DemGender'),
+    casout = list(name = 'treeModel1', replace = TRUE),
+    prune = TRUE
+ )
```

```
$ModelInfo
                          Descr        Value
1          Number of Tree Nodes 3.000000e+00
2       Max Number of Branches 2.000000e+00
3              Number of Levels 2.000000e+00
4              Number of Leaves 2.000000e+00
5                Number of Bins 2.000000e+01
6         Minimum Size of Leaves 7.656240e+05
7         Maximum Size of Leaves 9.233240e+05
8          Number of Variables 1.000000e+00
9   Confidence Level for Pruning 2.500000e-01
10  Number of Observations Used 1.688948e+06
11  Misclassification Error (%) 2.477163e+01

$OutputCasTables
                          casLib      Name Rows Columns
1     CASUSERHDFS(username) treeModel1    3      24
```

```
> output1 <- defCasTable(conn, 'treeModel1')
> output1[,c('_TreeLevel_', '_NodeID_', '_Parent_', '_ParentName_',
+            '_NodeType_', '_PBName0_',
+            '_PBName1_')]
```

```
  _TreeLevel_  _NodeID_  _Parent_  _ParentName_  _NodeType_  _PBName0_
1           0         0        -1                          1
2           1         1         0     DemGender           3          M
3           1         2         0     DemGender           3          F
  _PBName1_
1
2          U
3
```

Because the prune option is set to TRUE, Node 3 and Node 4 are pruned back according to the pruning criterion. The default pruning criterion is a modified C4.5 method, which estimates the error rate for validation data that is based on the classification error rate of the training data. The pruning criterion is not generated by the decisiontree.dtreeTrain action. However, you can use the information gain output that is generated by the tree model without pruning in order to understand why this split is not significant (the information gain is 0.0129) compared to the first split (the information gain is 0.0477).

Pruning a decision tree is used to avoid model overfitting when you use the decision tree to score new data (validation data). The C4.5 method does not require validation data in order to prune a tree. If you do have holdout data, you can use the decisiontree.dtreePrune action to prune a given tree model.

```
cas.decisiontree.dtreePrune(
  conn,
  table = 'your_validation_data',
  modelTable = 'treeModel1',
  casout = list(name = 'pruned_tree')
)
```

Next, let's add more variables to the decision tree models.

```
> varlist <- c('DemGender', 'DemHomeowner', 'DemAgeGroup', 'DemCluster',
+              'DemReg', 'DemTVReg', 'DemFlag1', 'DemFlag2', 'DemFlag3',
+              'DemFlag4', 'DemFlag5', 'DemFlag6', 'DemFlag7', 'DemFlag8',
+              'PromClass')
> cas.decisiontree.dtreeTrain(
+   organics,
+   target = 'TargetBuy',
+   inputs = varlist,
+   casout = list(name = 'treeModel2', replace = TRUE)
+ )
```

```
$ModelInfo
                          Descr      Value
1          Number of Tree Nodes 4.500000e+01
2          Max Number of Branches 2.000000e+00
3              Number of Levels 6.000000e+00
4              Number of Leaves 2.300000e+01
5                Number of Bins 2.000000e+01
6          Minimum Size of Leaves 7.600000e+01
7          Maximum Size of Leaves 5.972840e+05
8           Number of Variables 1.500000e+01
9  Confidence Level for Pruning 2.500000e-01
10  Number of Observations Used 1.688948e+06
11  Misclassification Error (%) 2.388966e+01


$OutputCasTables
                    casLib      Name Rows Columns
1    CASUSERHDFS(username) treeModel2   27     130
```

The ModelInfo output in the preceding example indicates that a larger decision tree has been grown. The new tree has 45 nodes and 23 of them are terminal nodes (leaves). The tree has 6 levels and the smallest leaves have 76 observations. For data with 1.7 million observations, sometimes, we might not be interested in subsets of data with only 76 observations. You can actually control subset size by the leafsize parameter. You can set leafsize to 1000 in order to grow a smaller tree, as follows:

```
> cas.decisiontree.dtreeTrain(
+   organics,
+   target = 'TargetBuy',
+   inputs = varlist,
+   leafSize = 1000,
+   maxLevel = 4,
+   casout = list(name = 'treeModel2', replace = TRUE)
+ )
```

```
$ModelInfo
                            Descr        Value
1          Number of Tree Nodes 1.500000e+01
2        Max Number of Branches 2.000000e+00
3             Number of Levels  4.000000e+00
4             Number of Leaves  8.000000e+00
5              Number of Bins   2.000000e+01
6         Minimum Size of Leaves 1.216000e+03
7         Maximum Size of Leaves 8.891240e+05
8          Number of Variables  1.500000e+01
9   Confidence Level for Pruning 2.500000e-01
10   Number of Observations Used 1.688948e+06
11   Misclassification Error (%) 2.393466e+01


$OutputCasTables
                      casLib      Name Rows Columns
1      CASUSERHDFS(username) treeModel2   15     130
```

In the previous example, the new decision tree produced only 15 nodes and 8 of them are leaves. Using this tree model to score a new observation is straightforward. You first identify the leaf node that contains this observation, and use the sample event probability of the node as the estimated event probability for that observation. The CAS decisiontree action set provides a dtreeScore action for you to score a data set as well.

```
> cas.decisiontree.dtreeScore(
+    organics,
+    modelTable = 'treeModel2'
+ )
```

```
$ScoreInfo
                            Descr               Value
1 Number of Observations Read            1688948
2 Number of Observations Used            1688948
3 Misclassification Error (%)        23.934662287
```

The model option in decisiontree.dtreeScore points to the tree model that we trained and stored on the CAS server, and it uses the model to score the Organics CAS table. The dtreescore action in this example does not actually save the predictions to a new CAS table. You must use a casout option to save the predictions.

```
> result <- cas.decisiontree.dtreeScore(
+    organics,
+    modelTable = 'treeModel2',
+    casout = list(name = 'predicted', replace = TRUE)
+ )
> output3 <- defCasTable(conn, 'predicted')
> names(output3)
```

```
 [1] "_DT_PredName_"   "_DT_PredP_"      "_DT_PredLevel_"  "_LeafID_"
 [5] "_MissIt_"        "_NumNodes_"      "_NodeList0_"     "_NodeList1_"
 [9] "_NodeList2_"     "_NodeList3_"
```

```
> head(output3, n = 10L)
```

	_DT_PredName_	_DT_PredP_	_DT_PredLevel_	_LeafID_	_MissIt_	_NumNodes_
1	No	0.6677494	1	9	0	4
2	No	0.7894737	1	13	0	4
3	No	0.8403450	1	8	0	4
4	No	0.9236861	1	7	0	4
5	No	0.6677494	1	9	0	4
6	No	0.6677494	1	9	0	4
7	No	0.9236861	1	7	0	4
8	No	0.6677494	1	9	0	4
9	No	0.8403450	1	8	0	4
10	No	0.6677494	1	9	0	4

	NodeList0	_NodeList1_	_NodeList2_	_NodeList3_
1	0	1	4	9
2	0	2	6	13
3	0	1	3	8
4	0	1	3	7
5	0	1	4	9
6	0	1	4	9
7	0	1	3	7
8	0	1	4	9
9	0	1	3	8
10	0	1	4	9

Table 9.5 summarizes the columns that are generated as output by the decisiontree.dtreeScore action.

Table 9.5: Summary of Output Columns from decisiontree.dtreeScore Action

Output Column	Description
_DT_PredName_	The predicted value, which is the most frequent level of the leaf that the observation is assigned to.
_DT_PredP_	The predicted probability, which is equal to the sample frequency of the predicted value in the leaf that the observation is assigned to.
_DT_PredLeveL_	The index of the predicted value. If the response variable has k levels, this column takes values from $0, 1, .., k - 1$
LeafID	The node ID of the leaf that the observation is assigned to.
MissIt	The indicator for misclassification.
NumNodes _NodeList$K$_	_NumNodes_ indicates the depth of the leaf. The list of _NodeList$K$_ variables stores the path from the root node to the current leaf.

Random Forests, Gradient Boosting, and Neural Networks

The decisiontree action set also contains the actions for building random forest models and gradient boosting. Unlike decision trees, random forests and gradient boosting are machine-learning techniques that produce predictions that are based on an ensemble of trees. Gradient boosting models are usually based on a set of weak prediction models (small decision trees or even tree stumps). By contrast, random forests are usually based on a set of full grown trees (deep trees) on subsample data. Another major difference is that gradient boosting grows decision trees sequentially and a random forest grows decision trees in parallel. For more details about tree models, refer to [1] and [2] at the end of the chapter.

Both random forest models and gradient boosting are available in the decisiontree action set through three distinct actions that cover basic steps of a machine learning pipeline: model training, scoring, and delivery (score code generation).

```
> cas.builtins.loadActionSet(conn, 'decisiontree')
```

```
NOTE: Added action set 'decisiontree'.
NOTE: Information for action set 'decisiontree':
NOTE:     decisionTree
NOTE:         dtreeTrain - Trains a decision tree
NOTE:         dtreeScore - Scores a table using a decision tree model
NOTE:         dtreeSplit - Splits decision tree nodes
NOTE:         dtreePrune - Prune a decision tree
NOTE:         dtreeMerge - Merges decision tree nodes
NOTE:         dtreeCode - Generates DATA step scoring code from a decision
                  tree model
NOTE:         forestTrain - Trains a forest
NOTE:         forestScore - Scores a table using a forest model
NOTE:         forestCode - Generates DATA step scoring code from a forest
                  model
NOTE:         gbtreeTrain - Trains a gradient boosting tree
NOTE:         gbtreeScore - Scores a table using a gradient boosting tree
                  model
NOTE:         gbtreeCode - Generates DATA step scoring code from a gradient
                  boosting tree model
```

Random Forests

Let's first train a simple random forest model using the Organics data set to predict the probability of buying organic food.

```
> varlist <- c('DemGender', 'DemHomeowner', 'DemAgeGroup',
+               'DemCluster', 'DemReg', 'DemTVReg', 'DemFlag1',
+               'DemFlag2', 'DemFlag3', 'DemFlag4', 'DemFlag5',
+               'DemFlag6', 'DemFlag7', 'DemFlag8', 'PromClass')
> cas.decisiontree.forestTrain(
+   organics,
+   target = 'TargetBuy',
+   inputs = varlist,
+   casout = list(name = 'forest1', replace = TRUE)
+ )
```

```
$ModelInfo
                            Descr       Value
1                 Number of Trees    50.00000
2   Number of Selected Variables (M)    4.00000
3             Random Number Seed     0.00000
4          Bootstrap Percentage (%)   63.21206
5                  Number of Bins    20.00000
6             Number of Variables    15.00000
7       Confidence Level for Pruning    0.25000
8         Max Number of Tree Nodes    59.00000
9         Min Number of Tree Nodes    31.00000
10        Max Number of Branches     2.00000
11        Min Number of Branches     2.00000
12          Max Number of Levels     6.00000
```

```
13              Min Number of Levels      6.00000
14              Max Number of Leaves     30.00000
15              Min Number of Leaves     16.00000
16          Maximum Size of Leaves 834199.00000
17          Minimum Size of Leaves     13.00000
18              Out-of-Bag MCR (%)            NaN

$OutputCasTables
                    casLib    Name Rows Columns
1      CASUSERHDFS(username)forest1 2224     132
```

The decisiontree.forestTrain action returns two result tables to the client: ModelInfo and OutputCasTables. The first table contains parameters that define the random forest, parameters that define each individual tree, and tree statistics, such as the minimum and maximum number of branches and levels. A random forest model is an ensemble of homogenous trees, with each tree growing on a different subset of the data (usually from bootstrap sampling). Therefore, the size and depth of the trees might be different from each other even though the tree parameters that you define are the same. The idea of a random forest is to grow a deep tree on each subsample in order to produce "perfect" predictions for the local data (low bias but high variance) and then use the ensemble technique to reduce the overall variance.

Some key parameters that define a random forest model are listed in Table 9.6. The decisiontree.forestTrain action also enables you to configure the individual trees, and the parameters are identical to those in the dtreetrain action. In general, you must grow deep trees for each bootstrap sample, and pruning is often unnecessary.

Table 9.6: Summary of Important Parameters for Random Forests

Parameter	Description
Ntree	The number of trees in the forest ensemble, which is 50, by default.
M	The number of input variables to consider for splitting on a node. The variables are selected at random from the input variables. By default, random forests use the square root of the number of input variables that are used, rounded up to the nearest integer.
vote='majority'	Uses majority voting to collect the individual trees into an ensemble. This is the default ensemble for classification models.
vote='prob'	Uses the average of predicted probabilities or values to collect the individual trees into an ensemble.
seed	The seed that is used for the random number generator in bootstrapping.
Bootstrap	Specifies the fraction of the data for the bootstrap sample. The default value is 0.63212055882.
Oob	The Boolean value to control whether the out-of-bag error is computed when building a random forest.

Random forest models are also commonly used in variable selection, which is usually determined by the variable importance of the predictors in training the forest model. The importance of a predictor to the target variable is a measure of its overall contribution to all the individual trees. In the decisiontree.forestTrain action, this contribution is defined as the total Gini reduction from all of the splits that use this predictor.

```
> result <- cas.decisiontree.forestTrain(
+    organics,
```

```
+    target = 'TargetBuy',
+    inputs = varlist,
+    varimp = TRUE,
+    casout = list(name = 'forest1', replace = TRUE)
+ )
> result['DTreeVarImpInfo']
```

```
$DTreeVarImpInfo
         Variable    Importance          Std
1        DemGender 15974.737141 8420.9397567
2      DemAgeGroup  7495.273449 2656.6087638
3        PromClass  1928.666310 1252.6262471
4       DemCluster   272.596220   76.1867226
5         DemTVReg   264.921498   90.4366253
6         DemFlag6   215.282293  318.5986969
7         DemFlag2   166.367985  245.2919257
8         DemFlag1   147.231080  180.4746655
9           DemReg   111.494411   51.7509476
10        DemFlag7    59.896469   39.0903619
11        DemFlag4    13.944902   29.4399686
12        DemFlag8    11.429714   29.4631055
13        DemFlag5     5.661628   43.5004329
14        DemFlag3     4.543302   19.0584222
15     DemHomeowner    0.364144    0.8927289
```

The decisiontree.forestTrain action also produces an OutputCasTables result table, which contains the name of the CAS table that stores the actual forest model. The CAS table, which is stored on the CAS server, describes all the individual trees. Each row of this CAS table contains the information about a single node in an individual tree. When you have a large of number of trees, this table can be large, and therefore, is stored on the CAS server instead. In the preceding example, the random forest model contains a total of 50 trees and 2278 nodes that are being returned to the client.

```
> result['OutputCasTables']
```

```
$OutputCasTables
                    casLib    Name Rows Columns
1     CASUSERHDFS(username) forest1 2254     132
```

To score the training data or the holdout data using the random forest model, you can use the decisiontree.forestScore action.

```
cas.decisiontree.forestScore(
  organics,
  modelTable = 'forest1',
  casout = list(name = 'scored_output', replace = TRUE)
)
```

Gradient Boosting

Unlike the random forest model, gradient boosting grows trees sequentially, whereby each tree is grown based on the residuals from the previous tree. The following example shows how to build a gradient boosting model using the same target and predictors, as in the random forest model.

```
> varlist <- c('DemGender', 'DemHomeowner', 'DemAgeGroup', 'DemCluster',
+              'DemReg', 'DemTVReg', 'DemFlag1', 'DemFlag2', 'DemFlag3',
```

```
+                 'DemFlag4', 'DemFlag5', 'DemFlag6', 'DemFlag7', 'DemFlag8',
+                 'PromClass')
> cas.decisiontree.gbtreeTrain(
+    organics,
+    target = 'TargetBuy',
+    inputs = varlist,
+    casout = list(name = 'gbtree1', replace = TRUE)
+ )
```

```
$ModelInfo
                                    Descr     Value
1                          Number of Trees     50.0
2                             Distribution      2.0
3                            Learning Rate      0.1
4                         Subsampling Rate      0.5
5      Number of Selected Variables (M)       15.0
6                           Number of Bins     20.0
7                      Number of Variables     15.0
8             Max Number of Tree Nodes        63.0
9             Min Number of Tree Nodes        61.0
10              Max Number of Branches         2.0
11              Min Number of Branches         2.0
12                Max Number of Levels         6.0
13                Min Number of Levels         6.0
14                Max Number of Leaves        32.0
15                Min Number of Leaves        31.0
16            Maximum Size of Leaves     426211.0
17            Minimum Size of Leaves         30.0
18                  Random Number Seed         0.0


$OutputCasTables
                       casLib    Name Rows Columns
1      CASUSERHDFS(username) gbtree1 3148     121
```

Table 9.7 lists the key parameters to grow a sequence of trees in a gradient boosting model. The decisiontree.gbtreeTrain action also enables you configure the individual trees. These parameters are identical to those in the dtreetrain action.

Table 9.7: Summary of Important Parameters for Gradient Boosting Trees

Parameter	Description
Ntree	The number of trees in the gradient boosting model. The default is 50.
M	The number of input variables to consider for splitting on a node. The variables are selected at random from the input variables. By default, all input variables are used.
seed	The seed for the random number generator for sampling.
subsamplerate	The fraction of the subsample data to build each tree. The default is 0.5.
distribution	The type of gradient boosting tree to build. The value 0 is used for a regression tree, and 1 is used for a binary classification tree.
lasso	The L1 norm regularization on prediction. The default is 0.
ridge	The L2 norm regularization on prediction. The default is 0.

To score the training data or the holdout data using the gradient boosting model, you can use the decisiontree.gbtreeScore action.

```
> cas.decisiontree.gbtreeScore(
+   organics,
+   modelTable = 'gbtree1',
+   casout = list(name = 'scored_output', replace = TRUE)
+ )
```

Neural Networks

Neural networks are machine learning models that derive hidden features (neuron) as nonlinear functions of linear combination of the predictors, and then model the target variable as a nonlinear function of the hidden features. Such nonlinear transformations are often called *activation functions* in neural networks. Figure 9.4 illustrates a simple neural network between three predictors and a binary target that takes levels 0 and 1. In this case, there are four hidden features. Each is a nonlinear function of the linear combination of the three predictors. The probabilities of target = 1 and target = 0 are modeled as a nonlinear function of the hidden features. In neutral networks, such hidden features are called *neurons,* and it is common to have more than one layer of hidden neurons.

Figure 9.4: The Structure of a Simple Neural Network between Three Predictors and a Binary Target

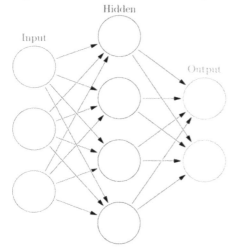

The neural network of related actions is available in the neuralnet action set. Let's load the action set and continue to use the Organics data set to build a simple neural network.

```
> cas.builtins.loadActionSet(conn, 'neuralnet')
```

```
NOTE: Added action set 'neuralnet'.
NOTE: Information for action set 'neuralnet':
NOTE:     neuralNet
NOTE:         annTrain - Trains an artificial neural network
NOTE:         annScore - Scores a table using an artificial neural network model
NOTE:         annCode - Generates DATA step scoring code from an artificial
                        neural network model
```

```
> result <- cas.neuralnet.annTrain(
+     organics,
+     target = 'TargetBuy',
+     inputs = c('DemAge','DemAffl','DemGender'),
+     casout = list(name = 'ann1', replace = TRUE),
+     hiddens = c(4,2),
+     maxIter = 500
+ )
> names(result)
```

```
[1] "ConvergenceStatus" "ModelInfo"         "OptIterHistory"
[4] "OutputCasTables"
```

In this case, we built a neural network with two layers. The first layer has four hidden neurons and the second layer has two hidden neurons. The maximum number of iterations for training the neural network is set to 500. The iteration history and the convergence status of the model are reported in OptIterHistory and ConvergenceStatus result tables, respectively. The ModelInfo result table contains basic information about the neural network model:

```
> result['ModelInfo']
```

```
$ModelInfo
                          Descr        Value
1                          Model    Neural Net
2    Number of Observations Used      1498264
3    Number of Observations Read      1688948
4       Target/Response Variable    TargetBuy
5                Number of Nodes           13
6          Number of Input Nodes            5
7         Number of Output Nodes            2
8         Number of Hidden Nodes            6
9        Number of Hidden Layers            2
10   Number of Weight Parameters           30
11     Number of Bias Parameters            8
12                  Architecture          MLP
13          Number of Neural Nets            1
14               Objective Value 1.7244440693
```

The default error function of the neuralnet.annTrain action is NORMAL for a continuous target or ENTROPY for a categorical target. The anntrain action also provides parameters for customizing the neural networks, such as the error function, the activation function, and the target activation function. Some key parameters of the neuralnet.anntrain action are listed in Table 9.8.

Table 9.8: Summary of Important Parameters for the neuralnet.anntrain Action

Parameter	Description
arch	Specifies the architecture of the network. arch = 'MLP': The standard multilayer perceptron network. arch ='GLIM': The neural network with no hidden layer. arch = 'DIRECT': The MLP network with additional direct links from input nodes to the target nodes.

Parameter	Description
errorfunc	Specifies the error function for training the neural network. ENTROPY is available for a categorical target. GAMMA, NORMAL, and POISSON are available for continuous targets.
targetact	Specifies the target activation function that links the hidden neurons at the last layer to the target nodes. LOGISTIC and SOFMAX (default) are available for categorical targets. EXP, IDENTITY, SIN, and TANH (default) are available for continuous targets.
act	Specifies the activation function that links the input nodes to the hidden neurons at the first layer, or neurons from one layer to the next layer. Available activation functions include EXP, IDENTITY, LOGISTIC, RECTIFIER, SIN, SOFTPLUS, and TANH (default).
targetcomb	Specifies the way to combine neurons in the target activation function. Linear combination (LINEAR) is the default. Other combinations are additive (ADD) and radial (RADIAL).
comb	Specifies the way to combine neurons or read as input in the activation functions. Linear combination (LINEAR) is the default. Other combinations are Additive (ADD) and radial (RADIAL).
ntries	The number of tries for random initial values of the weight parameters.
includebias	Indicates whether to include the intercept term (usually called bias) in the combination function. This parameter is ignored if an additive combination is used. Additive combinations are Combination (comb) and Target Combination (targetComb).

Similar to the tree models in CAS, a CAS table on the server can be used for storing the neural network model. This is convenient when you build a large neural network and it avoids the I/O traffic between the R client and the CAS server. To score a data set that uses a CAS table and that contains the neural network model, you can use the neuralnet.annScore action.

```
> cas.neuralnet.annScore(
+    organics,
+    modelTable = 'ann1',
+    casout = list(name = 'scored_output', replace = TRUE)
+ )
```

```
$OutputCasTables
                  casLib          Name     Rows Columns
1     CASUSERHDFS(username) scored_output 1688948       2

$ScoreInfo
                         Descr                      Value
1 Number of Observations Read                     1688948
2 Number of Observations Used                     1498264
3 Misclassification Error (%)                 18.377802577
```

Conclusion

In this chapter, we introduced several analytic models that are available on the CAS server for modeling categorical variables. This includes the logistic regression model in the regression action set, the tree family (decision tree, random forest, gradient boosting) in the decisiontree action set, and simple neural networks in the neuralnet action set.

References

[1] Friedman, Jerome, Trevor Hastie, and Robert Tibshirani. 2001. *The Elements of Statistical Learning.* Vol. 1. Springer, Berlin: Springer Series in Statistics.

[2] Breiman, Leo. 2001. "Random Forests." *Machine Learning* 45.1:5-32.

Chapter 10: Advanced Topics

Overview

There are a few topics that aren't necessary to use SWAT with CAS, but understanding them can help you in situations that you might come across. We'll look at the different network interfaces used by CAS, different styles of handling CAS action responses, and other miscellaneous topics.

Binary versus REST Interfaces

We mentioned in earlier chapters that there are two interfaces supported: binary and REST. The binary interface is a proprietary protocol supported on a limited number of platforms. This type of communication requires C extensions bundled with the SWAT installer on supported platforms. The REST interface allows SWAT to talk to the CAS server using the standard HTTP or HTTPS protocols. When using the REST interface, the SWAT client relies only on pure R code, so it can be used on any platform that can run R. We'll discuss the details of each in the following sections.

The Binary Interface

A diagram of the communication over the binary interface is shown in Figure 10.1.

Figure 10.1: Communication Over the Binary Interface

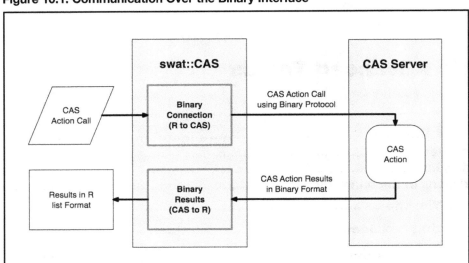

In this process, CAS action calls from the R code are converted to a binary parameter structure that is transferred to CAS using a proprietary protocol. On the server side, the parameter structure is unpacked and the action is executed. When the action finishes, the results are returned to the client in a binary packed form. The client then converts the binary responses back to R objects.

This form of communication is very fast and efficient, but it does require an R C extension and supporting libraries from SAS (the SAS TK framework) to run on the client. Because of the performance aspects of this protocol, it is recommended that you use it if it is available for your platform. When you make a connection to CAS, the SWAT client attempts to detect which type of port you are connecting to. The default is binary, but you can explicitly specify the protocol, as well as using the protocol parameter to the CAS object.

```
> binconn <- swat::CAS(cashost, casport, protocol='cas')
```

Let's look at the REST interface now.

The REST Interface

Figure 10.2 shows the process used in communications with CAS using the REST interface.

Figure 10.2: Communication Over the REST Interface

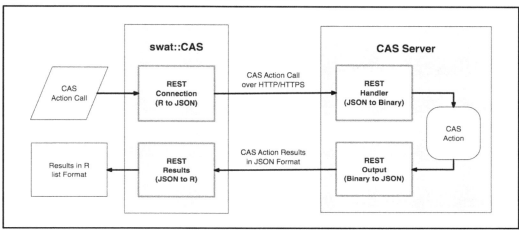

In this process, the CAS action parameters from R are converted to JSON (JavaScript Object Notation) format. The JSON parameters are sent to CAS using either the HTTP or HTTPS protocol. On the server side, the JSON format is converted to the binary format needed by the CAS action and the action is executed. The results, in binary form, are then converted to JSON format and sent back to the client. The client must then convert the JSON format into R objects to be returned in the R list object.

As you can see, there is much more overhead in using the REST interface. However, it also has the benefit of using pure R code, so it is more portable than using the binary interface.

To explicitly specify the REST interface when making a connection, you set the protocol option to 'http' or 'https' (depending on which protocol your server is configured to use).

```
> conn <- swat::CAS(cashost, casrestport, protocol='http')
```

The Pros and Cons of Each Interface

We'll summarize the pros and cons of each interface to make it easier for you to decide which one is best for your situation.

Binary (CAS) Interface

Pros

- Fast and efficient; not as many data conversions
- More authentication methods supported
- Supports custom data loaders using data message handlers
- The required SAS TK system also includes support for SAS data formats

Cons

- Not supported on all platforms

- The download size is much larger

REST Interface

Pros

- Uses standard HTTP/HTTPS protocols

- Uses pure R code, so it runs anywhere R is available

- Smaller download size

Cons

- More overhead due to conversions to and from binary formats

- Less efficient communication

- Data message handlers are not supported

Result Processing Workflows

Throughout this book, we have been calling actions using the methods on CAS connection objects and CASTable objects. Figure 10.3 shows the process used.

Figure 10.3: Process of Calling a CAS Action

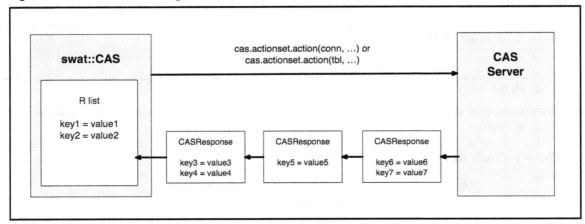

In this process, the CAS action is called on a CAS connection object or a CASTable object. The CAS server executes the action and returns multiple responses back to the client. Each response contains one or more key/value pairs of results. The SWAT client processes each response from the server and combines them all into a single list object. Once all responses have been received, the combined results are returned.

Connecting to Existing Sessions

While you can only run one action in a session at a time, it is possible to connect to existing sessions multiple times. This can be handy if you ran some actions while at work and you want to continue your work from home when they are finished. Connecting to an existing session is fairly easy, you just need to know the session ID. Each R connection is tagged with a unique session ID. You can see it when you display the string representation of a CAS connection object.

```
> conn
CAS(hostname=server-name.mycompany.com, port=8777,
            username=username, protocol=http,
            session=c4bbd504-d4a7-134f-af62-9340b3dc8048)
```

When you want to connect to that session from another location, you can use the session.listSessions action to display all of the current sessions on the server. This lists all of the sessions on the server created by you.

```
> cas.session.listSessions(conn)
```

```
$Session
                        SessionName
1    Session:Thu Feb  8 14:10:50 2018
2    Session:Thu Feb  8 14:45:38 2018
                            UUID      State    Authentication
1 5f7993ad-6a5a-cb40-8632-26b8e457f893 Connected Active Directory
2 1d4f6b7e-d213-cb44-9a9e-a9a7c2e5194b Connected Active Directory
   Userid
1 username
2 username
```

We can now get the full session ID by copying the corresponding UUID value. Now that we have that, we can specify the UUID in the session parameter of a new connection.

```
> conn2 <- CAS(hostname = 'server-name.mycompany.com', port = 8777,
+    session = '5f7993ad-6a5a-cb40-8632-26b8e457f893')
```

Your new connection is now connected to the specified session. Also, since the sessions are language agnostic, you can connect to existing CAS sessions created from other language interfaces, such as Java or SAS.

Communicating Securely

If your server is configured to use TLS encryption either through the binary interface or using the REST interface through HTTPS, you will need to set the path to the file that contains the certificate chain to verify your connections. See the SAS documentation on the different methods for creating your certificate files for use with CAS. Configuring the client is done by setting a CAS_CLIENT_SSL_CA_LIST environment variable. The method for setting environment variables varies by platform (or even command shell on Linux). Below is the way to set the environment variable using the bash shell on Linux.

```
export CAS_CLIENT_SSL_CA_LIST='/path/to/cacerts/trustedcerts.pem'
```

On Windows, the relevant server certificate must be imported into the Windows certificate store. Locate the proper certificate on your server; it will typically be named vault-ca-crt, vault-vault_host_role-ca.crt, or httpproxy-httpproxy_host_role-ca.crt. Right-click the certificate and choose Install Certificate. Set the location to Local Machine and click Next. On the next screen, browse for the certificate store location and choose "Trusted Root Certificate Authorities". Click Next and Continue to finish.

Conclusion

With the information in this chapter, you should be able to make educated decisions on which interface (binary or REST) that you want to use. You learned how the result processing workflows work. You can also connect to existing sessions, and connect to CAS using encrypted communications. This knowledge should enable you to handle more advanced uses of CAS connections and workflows.

Index

Ready to take your SAS® and JMP® skills up a notch?

Be among the first to know about new books,
special events, and exclusive discounts.
support.sas.com/newbooks

Share your expertise. Write a book with SAS.
support.sas.com/publish

sas.com/books
for additional books and resources.

§.sas

THE POWER TO KNOW.

www.ingramcontent.com/pod-product-compliance
Lightning Source LLC
Chambersburg PA
CBHW080528060326
40690CB00022B/5067